Anja Mueller
Biomimetic Nanotechnology

Also of Interest

Biomimetics
A Molecular Perspective
Raz Jelinek, 2021
ISBN 978-3-11-070944-5, e-ISBN (PDF) 978-3-11-070949-0,
e-ISBN (EPUB) 978-3-11-070994-0

Bioelectrochemistry
Design and Applications of Biomaterials
Serge Cosnier (Ed.), 2019
ISBN 978-3-11-056898-1, e-ISBN (PDF) 978-3-11-057052-6,
e-ISBN (EPUB) 978-3-11-056926-1

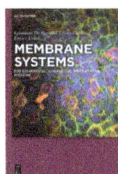

Membrane Systems
For Bioartificial Organs and Regenerative Medicine
De Bartolo, Curcio, Drioli (Eds.), 2017
ISBN 978-3-11-026798-3, e-ISBN (PDF) 978-3-11-026801-0,
e-ISBN (EPUB) 978-3-11-039088-9

Nanoscience and Nanotechnology
Advances and Developments in Nano-sized Materials
Van de Voorde (Ed.), 2018
ISBN 978-3-11-054720-7, e-ISBN (PDF) 978-3-11-054722-1,
e-ISBN (EPUB) 978-3-11-054729-0

Anja Mueller

Biomimetic Nanotechnology

Senses and Movement

2nd, Completely Revised Edition

DE GRUYTER

Author
Prof. Anja Mueller
Central Michigan University
Dept. of Chemistry
374 Dow
Mt. Pleasant, MI 48859
USA
muell1a@cmich.edu

ISBN 978-3-11-077918-9
e-ISBN (PDF) 978-3-11-077919-6
e-ISBN (EPUB) 978-3-11-077947-9

Library of Congress Control Number: 2022950726

Bibliographic information published by the Deutsche Nationalbibliothek
The Deutsche Nationalbibliothek lists this publication in the Deutsche Nationalbibliografie;
detailed bibliographic data are available on the Internet at http://dnb.dnb.de.

© 2023 Walter de Gruyter GmbH, Berlin/Boston
Cover image: Lan Zhang / iStock / Getty Images Plus
Typesetting: VTeX UAB, Lithuania
Printing and binding: CPI books GmbH, Leck

www.degruyter.com

Preface

This book is about human senses and movement, how they function on the nanoscale, and how they can be mimicked on the nanoscale by technology. It combines viewpoints and data from a variety of different fields: biology and chemistry, of course, but also physics and mechanical and electrical engineering. Since these fields are generally taught separately, many students may find this interdisciplinary presentation challenging. Therefore, I would recommend this book for senior undergraduate or early graduate students. While chemistry and biology are the most important fields in this book, students from any of the above-mentioned fields may read this book. It has been purposely written in simple language with explanations of the jargon used to make it accessible to a variety of fields.

The introduction presents the major points of each field needed to understand the content. Each student can choose the sections needed for their background. Each chapter starts by first looking at the molecular details of the sense or movement, so that it is clear what exactly is mimicked. The next part of each chapter will look at technology that uses some of the human molecules in nanotechnology. The last part of each chapter looks at nanotechnology that mimics the function of each sense or movement, not by using human molecules but chemistry, engineering, biology, and other sciences to achieve its goal.

Biomimetic nanotechnology is a large, quickly moving field. A lot of exciting new developments have been added to this edition. For the first time, research attempts to start with a nanoscale sensor, then to scale it up by self-assembly to a large-scale structure that includes many sensors, similar to what happens in human organs. This book is not a comprehensive compilation of all research on the subject; instead, I have chosen several examples for each chapter that stress the connection between the molecules in the human body and the human function of the senses and movement.

https://doi.org/10.1515/9783110779196-201

Acknowledgment

I would like to thank The Designlizard (www.designliz.art; andrea.mangold@design-liz.art) for the majority of the graphs that are not chemical structures. I would also like to thank Ajit Sharma for providing access to his book, Xu Zhang for help with micro-electronics, and Ajit Sharma and Ben Swarts for many interesting discussions on the topic.

https://doi.org/10.1515/9783110779196-202

Contents

List of abbreviations

AI	Artificial intelligence
AMP	Adenylate monophosphate
ATP	Adenylate triphosphate
CaM	Calmodulin
cAMP	cyclic Adenylate monophosphate
CCD	Charge-coupled device
CO	Carbon monoxide
CytP450	Cytochrome P450 enzyme
DAG	Diacyl glycerol
DG	Diacyl glycerol
DNA	Deoxyribonucleic acid
DPH	N'1,N'6-bis(3-(1-pyrrolyl) propanoyl) hexanedihydrazide
E	Young's modulus
ECM	Extracellular matrix
ER	Endoplasmic reticulum
$FADH_2$	Flavin adenine dinucleotide
FET	Field effect transistor
FRET	Fluorescence resonance energy transfer
GABA	γ-Aminobutyric acid
GDP	Guanine diphosphate
GLcNAc	N-Acetylglucosamine
GPC	Gel-permeation chromatography
GPCR	G-protein-coupled receptors
GTP	Guanine triphosphate
HPLC	High-pressure liquid chromatography
IC	Integrated circuit
IEC	Ion-exchange chromatography
IP_3	Inositol triphosphate
ITO	Indium tin oxide
LAPS	Light-addressable potentiometric sensor
LED	Light-emitting diode
MEA	Microelectrode array
MEMS	Micro-electrical-mechanical systems
MET	Mechanoelectrical transduction
MIP	Molecularly-imprinted particles
MOSFET	Metal-oxide semiconductor field-effect transistor
NADH	Nicotinamide adenine dinucleotide
NEMS	Nano-electrical-mechanical systems
NO	Nitric oxide
NOx	Nitrogen oxides, NO and NO_2
NS	Nervous system
ODP	Odorant-binding protein
ODR	Odorant receptor
PDMS	Polydimethylsiloxane
PKC	Protein kinase C
PLC	Phospholipase C
QCM	Quartz crystal microbalance

https://doi.org/10.1515/9783110779196-203

RGD   Arginine, glycine, aspartic acid
RNA   Ribonucleic acid
SAM   Self-assembled monolayer
SEC   Size-exclusion chromatography
SOx   Any mixture of lower sulfur oxides (S_nO), sulfur monoxide (SO), sulfur dioxide (SO_2), sulfur trioxide (SO_3), higher sulfur oxides (SO_3,SO_4, polymeric condensates), disulfur monoxide (S_2O), disulfur dioxide (S_2O_2)
T_g Glass transition temperature
TRP   Transient receptor potential ion channel
tRNA   transfer-RNA
UV   Ultraviolet
Vis   Visible

1 Introduction

This book is about human senses and movement, how they function on the nanoscale, and how they can be mimicked on the nanoscale by technology. To understand this subject, it is necessary to first know a little about what happens in the human body and what takes place at the nanoscale more generally. This introduction will help you by providing an overview on several topics. It will also include common measurement techniques and an introduction to sensors. Feel free to only look at the sections you need help with; each section should be understandable by itself. References will be provided if you want to look into a topic in more detail.

1.1 Introduction to Cell Biology and Cell Communication

Let's start with some basic physiology [1]. Skin, the biggest organ of the body, surrounds and protects the other organs, each with their separate function. The skeleton holds up all of the organs and cells, the brain coordinates reactions to our environment, plans and regulates movement, the lungs provide oxygen so that fuel can be burnt and energy for the body created, this fuel is collected and broken down in the stomach, but toxic compounds and waste are secreted via the kidneys. The liver produces whatever needs to be produced to make it all work, and the heart is responsible for the transport of nutrients throughout the body. Why does each function have its own organ?

The short answer is that for each organ to be able to function correctly it needs its own environment [1]. Each organ has different genes that are active and works with different chemical compounds in its cells, using a variety of reactions to reach its goals. And each organ, and the body itself, can only function in very controlled conditions within a narrow margin. These narrow margins are called homeostasis. The body has an overall homeostasis or set of requirements and conditions, but so does each organ and each individual cell.

To maintain those conditions, membranes envelop the body, each organ, and each cell. These border-enclosed spaces can then be controlled separately. But here is the problem with this approach: each of those separated spaces, including each cell, must still work together to create a functioning body. Humans have trillions and trillions of cells, and yet humans are not just a random heap of cells. The cells organize themselves into tissues, each tissue is ordered into organs, and each organ is combined to form a human. Humans have a hard time organizing several people; how can the body organize trillions of cells? And not only organize the cells into a static structure, but create an active, environment-responsive, moving and thinking human? The answer is: communication, communication, communication.

Before we get to the communication part, let us look at the structure of a cell and how it functions (Figure 1.1). The nucleus contains all information for the cell coded in the double-stranded deoxyribonucleic acid (DNA). The sequence of four different bases

https://doi.org/10.1515/9783110779196-001

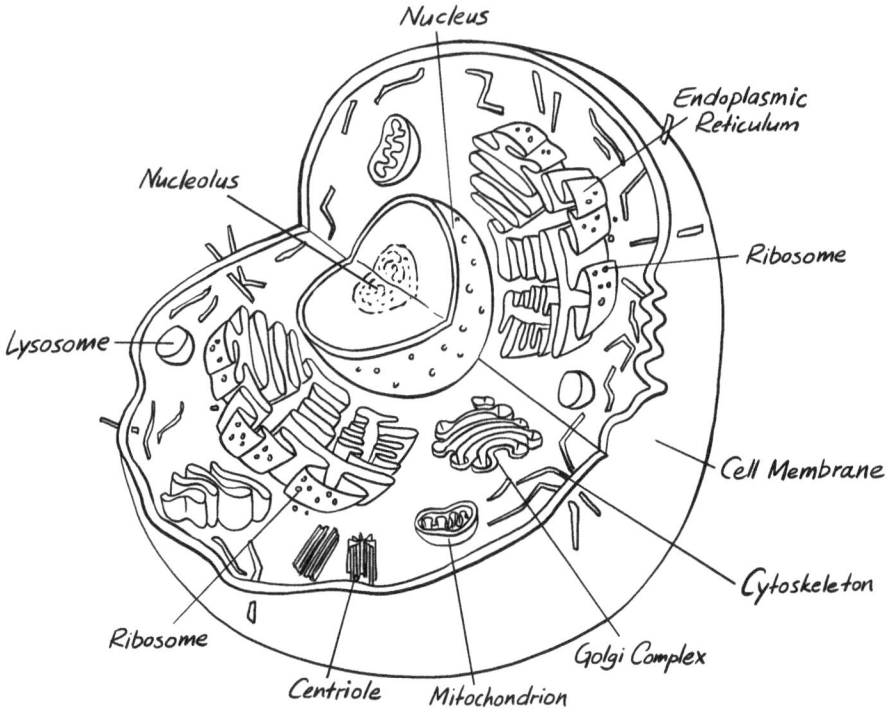

Figure 1.1: Cell Structure and Components.

in DNA gets translated into ribonucleic acid (RNA) and then proteins by ribosomes, and the proteins are distributed to their specific locations in the cell via the endoplasmic reticulum (ER) and the Golgi complex. Some proteins are structural, creating the different structures in the cell and human body. Most of the proteins are enzymes, i. e. they catalyze reactions and thus either construct needed compounds in the cell or break them down to generate energy (anabolism or catabolism, respectively). Some catalysts are also part of signal transduction pathways and thus part of cell communication.

The place where energy is generated is in the mitochondria. Lysosomes, vacuoles, and peroxisomes either ingest fuel or break down toxic compounds. The cytoskeleton is made from structural protein fibers and holds the shape of the cell and act as "roads" for organelle and other transport throughout the cell. The cell membrane is the barrier that allows for homeostasis. It contains a lot of proteins which are mostly channels that allow for very specific and regulated transport in and out of the cell. Other transport across the membranes operates via vesicles that are endocytosed or exocytosed. Other transmembrane proteins are designed to transfer signals. Some proteins on the outside of the cell membrane act as a "marker", or recognition element, for the cell, and some cause the cell to adhere to other cells or the extracellular matrix (ECM). ECM consists of fibers (collagen, elastin) and an amorphous matrix (proteoglycans, cell-binding adhesive glycoproteins, solutes, and water) which together form a cell-support. The mechanical strength of the tissue depends on both the strength of the cytoskeleton for each cell and the strength of the ECM. The ordering of cells into tissues is also regulated by the ECM.

Cells are not static structures but grow, develop and mature, and multiply. They also react to stimuli, outside forces and conditions, and might even commit suicide (apoptosis) if they are irreversibly damaged or are infected with a dangerous virus or bacterium. The centriole starts the process of cell division (mitosis) by making the spindle fibers that draw the two strands of the chromosomes (DNA pieces) apart. The cells can then form two nuclei, one for each cell, by copying the one strand, reforming the complete, double-stranded DNA chromosome and then forming a nuclear membrane around it; after that, the complete cell divides by forming a membrane between those nuclei and separating the identical cells from each other.

To modify a cell, specifically to modify or remove a specific protein in a cell, the method of genetic engineering was developed. It has been most widely used in bacteria and plants. The genetic information for a specific protein, its gene or piece of DNA sequence, is excised and modified. To be active in the cell, that piece of DNA needs to be attached to a promoter sequence that regulates when and how that information is used, then the DNA is reinserted into the cell. This method works most effectively if one function is based on only one gene, which is common in simple organisms such as bacteria but rare for higher organisms.

With this information on cell structure in mind, let us return to cell communication. The body has two major communication systems: the endocrine system that uses hormones as its signal and the neuronal system that uses neurotransmitters instead (Figure 1.2). Hormones are secreted into the blood stream and are widely distributed. They act on any tissue that happens to have receptors for that specific hormone. Hormones are generally signals for development, i. e., slower, more long-term processes such as the signal for a stem cell to develop into a new neuron in the brain, or signals that activate several organs, i. e. insulin, which activates several organs so that food can be broken down and converted into energy.

Neurotransmitters will be released into a very small space between two nerve cells and thus will act only on that subsequent neuron. Release and uptake are fast, and the neurotransmitter will be destroyed immediately. So these signals are used to react quickly to the environment or analyze and transmit information in the brain stemming from the continuous input of our senses.

In either communication system, though, the signal arrives outside of the cell that is supposed to use that information. So how does the signal on the **outside** of the cell effect a change **inside** of that cell, given that the cell membrane poses such a formidable barrier? Here, the specific channels or other transmembrane proteins come into play. The structures and mechanisms of these channels will be discussed in more detail in the next chapter (see Section 1.2). The signal binds to a channel or a transmembrane protein on the outside of the cell. That binding event changes the three-dimensional structure of that protein so that an enzyme at the inside of the cells gets activated to catalyze a

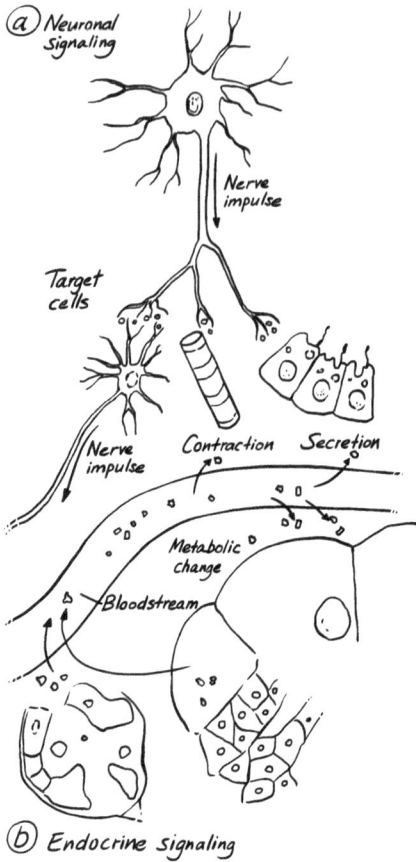

(a) Neuronal Signaling

Nerve impulse

Target cells

Nerve impulse *Contraction* *Secretion*

Metabolic change

Bloodstream

(b) Endocrine signaling

Figure 1.2: Signaling between cells; (a) Neuronal signaling; (b) Endocrine signaling.

reaction. The product of that reaction then binds to another enzyme, which activates it to perform another reaction, whose product binds and activates another enzyme, and so on. This process is called a signal cascade.

Why does a signal cascade take place, instead of just one single reaction? Doesn't such a complex process create a greater risk of something going wrong? In fact, the opposite is the case: Each of the steps in the signal cascade can be regulated, thus allowing the signal effect to be tightly controlled. Also, a signal cascade creates many signals for a lot of proteins in a short time, and thus creates signal amplification, which allows for fast, coordinated action. For example, within a muscle, a large number of heads of myosin have to be activated at the same time, otherwise only a few muscle fibers would contract instead of the complete muscle.

We will look at two common signal transduction pathways as examples. The first is the adenylate cyclase pathway (Figure 1.3). The transmembrane receptor is associated with a G-protein (a protein that hydrolyses guanine triphosphate (GTP) to guanosine diphosphate (GDP)). These complexes are called G-protein-coupled receptors, or GPCR.

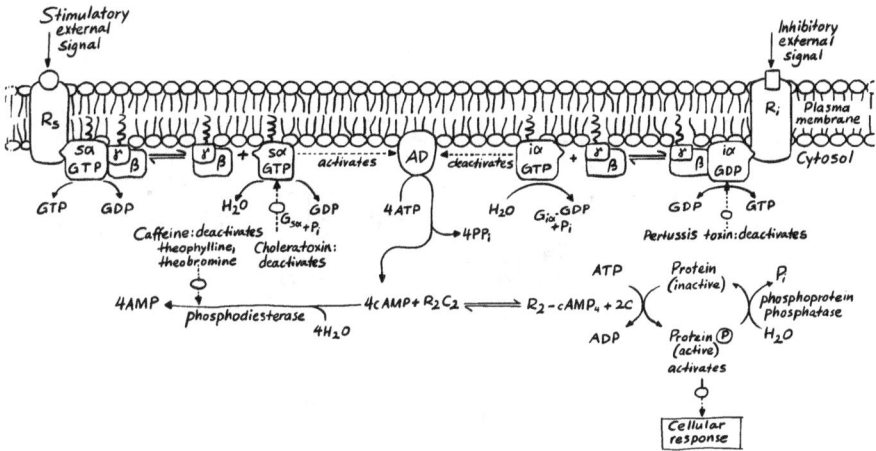

Figure 1.3: Adenylate-cyclase signal transduction pathway.

The G-protein is phosphorylated and in turn phosphorylates the enzyme adenylate cyclase. The thus activated adenylate cyclase takes adenylate monophosphate (AMP) and cyclizes it to cAMP; cAMP is an important signal in the cell that can bind and activate many kinases. Kinases are enzymes that can phosphorylate other enzymes and thus activate or inactivate enzymes, thereby allowing or barring specific reactions. Each of the enzymes in the pathway can be regulated either by binding a different compound into the active site of the enzyme or by binding to a different site in the enzyme, which is called an allosteric site. When binding to an allosteric site, the enzyme can still react with its compound in the active site, but the binding strength and thus the reaction is modulated (i. e., it slows down).

Another example of a common signal transduction pathway is the phospholipase C pathway (Figure 1.4). Again, the signal binds to a transmembrane protein that is associated with a G-protein, a GPCR. In this case, the G-protein activates the enzyme phospholipase C, which generates two signals, inositol triphosphate (IP_3) and diacyl glycerol. Diacyl glycerol in turn activates kinases as we have seen before. On the other hand, IP_3 activates ion channels in the ER that release calcium (Ca^{2+}) ions from storage. Ca^{2+} ions are another important internal signal that in turn can activate another set of kinases, but it can also control muscle contractions. Most of the senses also use signal transduction pathways like these to react to outside signals.

To understand the signaling process at the nanoscale, it is important to look at the nanoscale itself and the structures of some of the molecules involved, such as GPCR, in more detail. This will allow us to then use and control these molecules in nanotechnology.

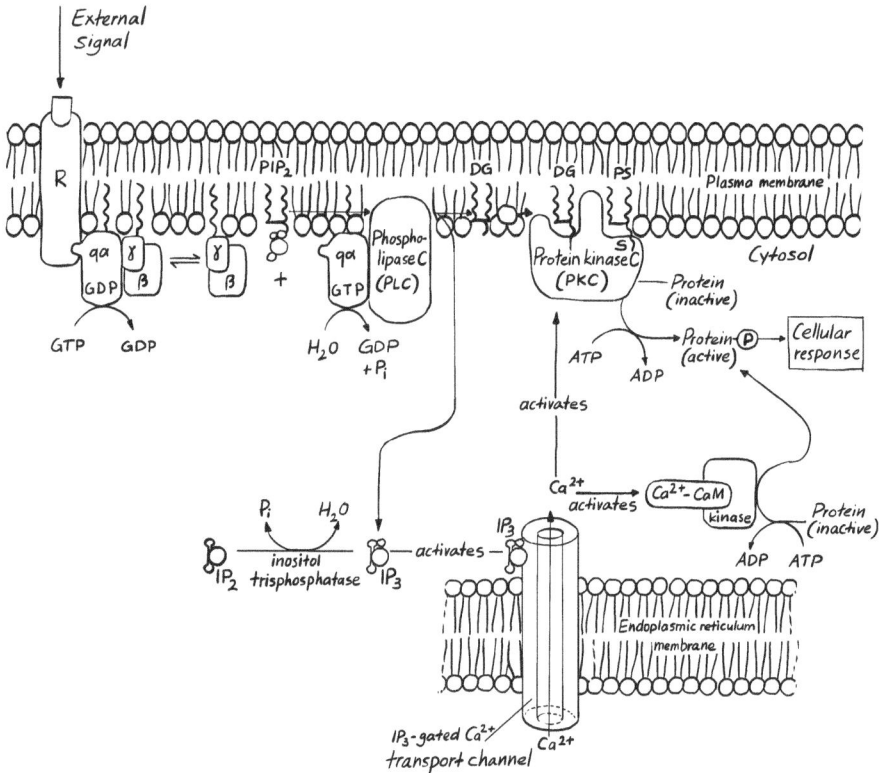

Figure 1.4: Phospholipase C signal transduction pathway.

1.2 Nanoscale Actors and Their Properties

Nanoscale. The nanoscale is generally defined as anything within the length scale of one to a hundred nanometers, a billionth of a meter (Figure 1.5). A nanometer is only ten times larger than an Angstrom, which is the size of a small single atom. There is technology at this scale at this point, but most small technology officially resides in the microscale, with some parts operating at the nanoscale. In these cases, the name nanoscale will be used more loosely.

Why is scale so important? It turns out that the properties of materials change when comparing the nanoscale with the macroscale, or bulk, materials [2]. Looking simply at geometry, the volume or bulk of a material grows with the cube of the length, the area only with the square of the length. That means that in a bulk material, there is mostly bulk (hence the name) and very little surface, while the nanoscale material is the opposite, a lot of surface and very little bulk. Why is this important? Atoms always want to bond with all the orbitals they have, which is, when they are at their lowest energy, the most stable and the least reactive. Atoms on a surface, though, only interact with the orbitals that face the bulk, the orbitals facing the outside have nothing to bond

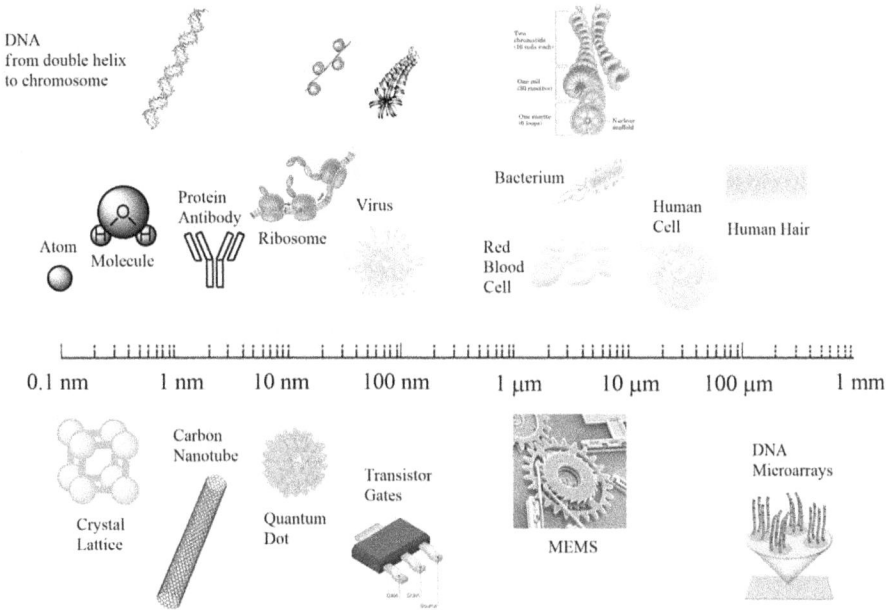

DNA
from double helix
to chromosome

Atom

Molecule

Protein
Antibody

Ribosome

Virus

Bacterium

Red
Blood
Cell

Human
Cell

Human Hair

0.1 nm   1 nm   10 nm   100 nm   1 μm   10 μm   100 μm   1 mm

Crystal
Lattice

Carbon
Nanotube

Quantum
Dot

Transistor
Gates

MEMS

DNA
Microarrays

Figure 1.5: The scale of small things, nanoscale and microscale.

with. Thus the surface atoms are higher in energy and more reactive. This makes a large, bulk material generally unreactive, and nanoscale material highly reactive. This is also true for possible reactions on the surface of a material as well as interactions of the surface with its environment.

This principle leads to some interesting phenomena. To give an example on a still more human scale, a human will break through the surface of water when stepping on it, but a water strider cannot, since the strider cannot break the intermolecular forces on the surface of water that make up its surface tension [2]. The smaller the surface, the larger the effect that surface tension has. That also changes the flow properties through small-diameter channels, which will become important for microfluidic devices, since the intermolecular forces, i. e., viscosity, will become predominant.

Keeping nanoscale properties in mind, let us look at the structure of the molecules that become important in the body and in nanotechnology. The most important materials in a cell are made from either lipids, nucleotides, sugars, or amino acids. We'll look at all of these materials in more detail, starting with lipids.

Structure and function of molecules – lipids and lipid membranes. The most common lipids are made from fatty acids that are attached to a head group (Figure 1.6). The fatty acid is by itself amphiphilic, i. e. it has a hydrophilic head and a long, hydrophobic tail. In water, therefore, fatty acids will aggregate in such a way that they form a phase where the hydrophilic heads face the water and the hydrophobic tails face away from it. This phase, though, is not a solid phase like a salt crystal, but a liquid crystal phase. A liquid

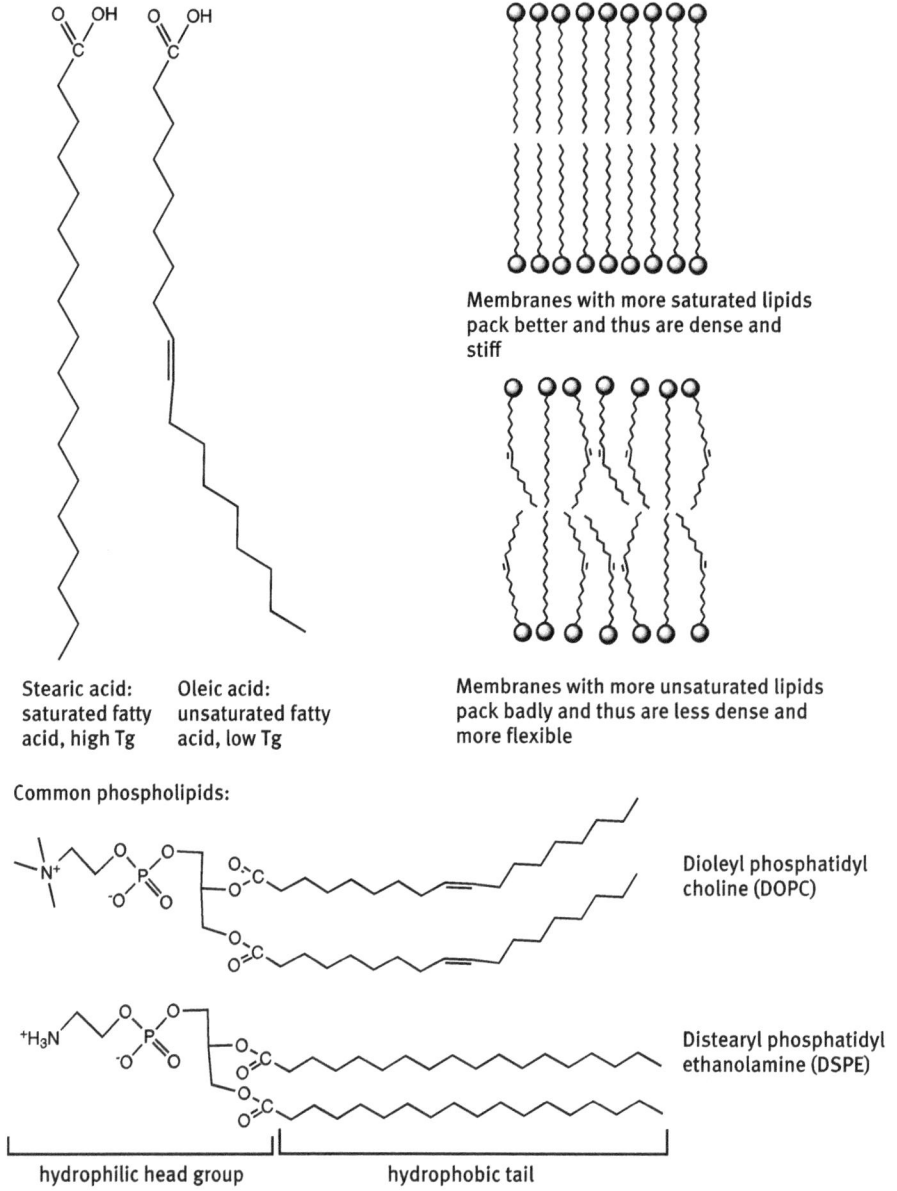

Membranes with more saturated lipids pack better and thus are dense and stiff

Membranes with more unsaturated lipids pack badly and thus are less dense and more flexible

Stearic acid: saturated fatty acid, high Tg

Oleic acid: unsaturated fatty acid, low Tg

Common phospholipids:

Dioleyl phosphatidyl choline (DOPC)

Distearyl phosphatidyl ethanolamine (DSPE)

hydrophilic head group   hydrophobic tail

Figure 1.6: The structures of the membrane lipids (mostly phospholipids) and their fatty acids determine the mechanical properties of the membrane.

crystal phase maintains its shape and order at all times, while the individual fatty acids actually change places throughout the structure, thus behaving like a moving liquid. A liquid crystal phase can still "melt", i. e., disintegrate, but the phase transition is not a melting point, but a glass transition temperature, T_g. The more ordered the phase is, as with straight, saturated fatty acids, the higher the T_g. The less ordered the phase is, as with bent, unsaturated fatty acids, the lower the T_g. A layer of fatty acids can also be formed at the surface of water and then transferred onto a solid surface. These fatty acid single layers are called self-assembled monolayers (SAM). They are also the reason why amphiphiles are called surfactants (from "<u>surf</u>ace <u>act</u>ive", or moving to the surface).

The most common lipids have their hydrophilic head attached to a hydrophilic head group, thus overall still ending up with an amphiphilic molecule. Here, we will specifically mention phospholipids, or phosphoglycerides (Figure 1.6), since they are the main lipids that make up cell membranes and the most commonly used lipids in nanotechnology. The hydrophilic head group is phosphoglycerol, the hydrophobic tail contains two fatty acid tails, either saturated or unsaturated, with somewhat different lengths.

Do to their amphiphilicity, lipids form liquid crystals in water (Figure 1.7). Depending on the shape of the lipid a bilayer can form, which is the basis of all lipid bilayer cell membranes. But differently-shaped liquids can also form single-layer or double-layer spheres, called micelles or liposomes, respectively. All of them have been used in nanotechnology. The mechanical properties of lipid membranes can be manipulated by changing the composition of the bilayers. A membrane containing more saturated lipids will have a higher T_g, and thus will be stiffer, than a membrane with more unsaturated lipids. The large, stiff cholesterol molecules further stiffen membranes, as do proteins, especially transmembrane proteins, which are proteins that are long enough to cross the membrane completely.

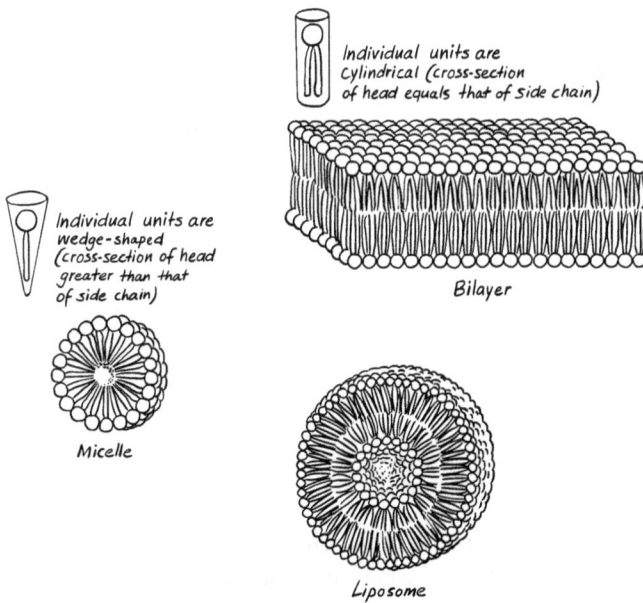

Individual units are cylindrical (cross-section of head equals that of side chain)

Bilayer

Individual units are wedge-shaped (cross-section of head greater than that of side chain)

Micelle

Liposome

Figure 1.7: Common liquid crystal structures for phospholipids in water.

Structure and function of molecules – sugars and polysaccharides. Another important polymer in the body is made out of sugars. A polymer is a molecule that is made from many ("poly") repeating units ("mer"). They can be arranged into a long chain or into a branched structure. There are a variety of different sugars in the body; they are either used for energy or for structural polymers. One of the most important sugars in the body is glucose (Figure 1.8). Glucose is the sugar that is fed directly into glycolysis, one of the major energy-generating pathways, and which is regulated by the hormone insulin. Derivatives of glucose are also common structural sugars making up, for example, our tendons. Additionally, glycogen is a partially branched polymer out of glucose that is the major energy storage molecule besides lipids. Due to branching and the α-linkages, glycogen is a soft material (similar to but even softer than starch). It is stored in granules in cells (especially muscle cells) and enzymatically broken down when needed.

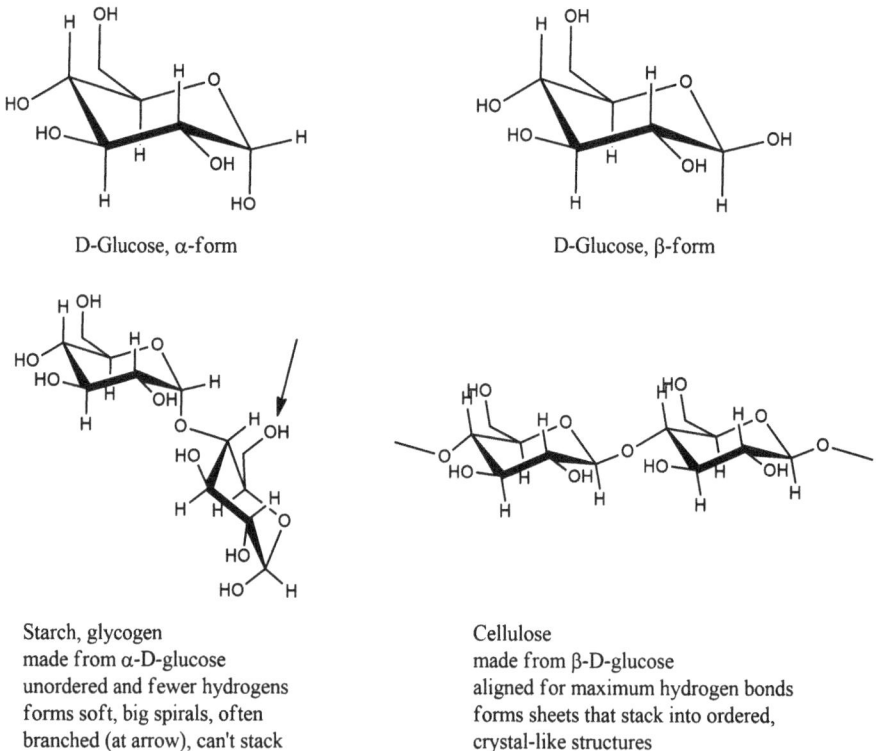

D-Glucose, α-form

D-Glucose, β-form

Starch, glycogen
made from α-D-glucose
unordered and fewer hydrogens
forms soft, big spirals, often
branched (at arrow), can't stack

Cellulose
made from β-D-glucose
aligned for maximum hydrogen bonds
forms sheets that stack into ordered,
crystal-like structures

Figure 1.8: Structure of glucose and polymers of glucose (examples for polysaccharides).

Structure and function of molecules – nucleotides and DNA and RNA. Another group of materials in the body are the nucleotides or the polymers made from nucleotides, DNA and RNA. In the case of DNA and RNA, the "mers" of the polymer are nucleotides, which can carry 4 different bases, and in the sequence of these bases the information of genes

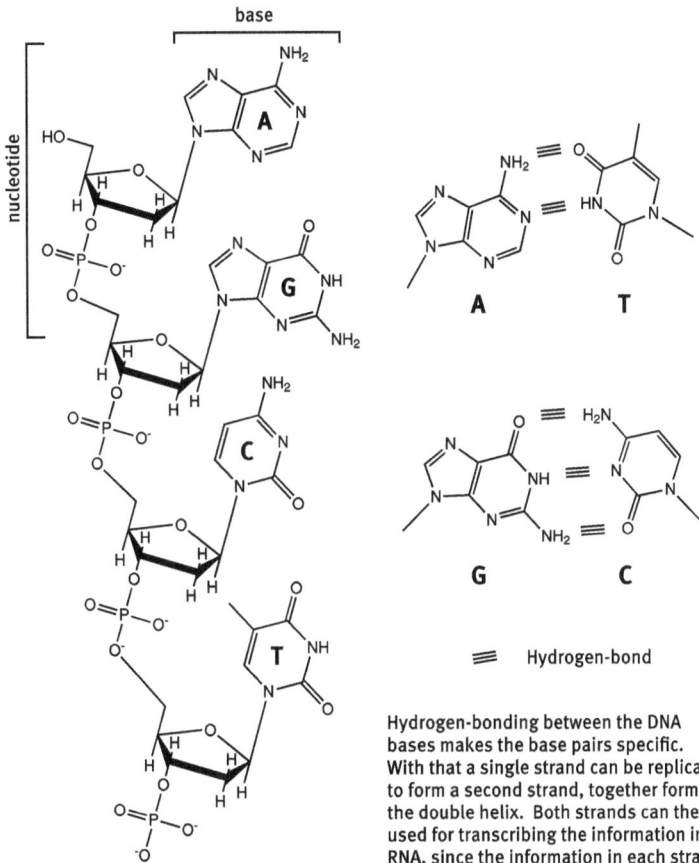

A single strand of DNA with its four different bases: adenenine (A), guanine (G), cytosine (C), and thymine (T).

Hydrogen-bonding between the DNA bases makes the base pairs specific. With that a single strand can be replicated to form a second strand, together forming the double helix. Both strands can then used for transcribing the information into RNA, since the information in each strand is the same.

Figure 1.9: Structure of DNA and the base pairs of DNA.

is encoded (Figure 1.9). Additionally, a base can interact with a base on another DNA strand with specific hydrogen bonds, which are strong enough that another structure results, the famous double helix. The specificity and strength of the hydrogen bonds is such that it can be used to translate the DNA information into proteins or create more DNA that is exactly the same in mitosis (see Section 1.1).

Structure and function of molecules – amino acids, peptides and proteins. The other important polymers in the body are peptides or proteins, which are made out of amino acids. Here, the specificity due to hydrogen bonding and other intermolecular forces continues: Proteins are the specific polymers that DNA information is translated into. Basically, specific hydrogen bonds of DNA are transcribed into specific hydrogen bonds of RNA, which then are translated via specific hydrogen bonds to transfer-RNA (tRNA),

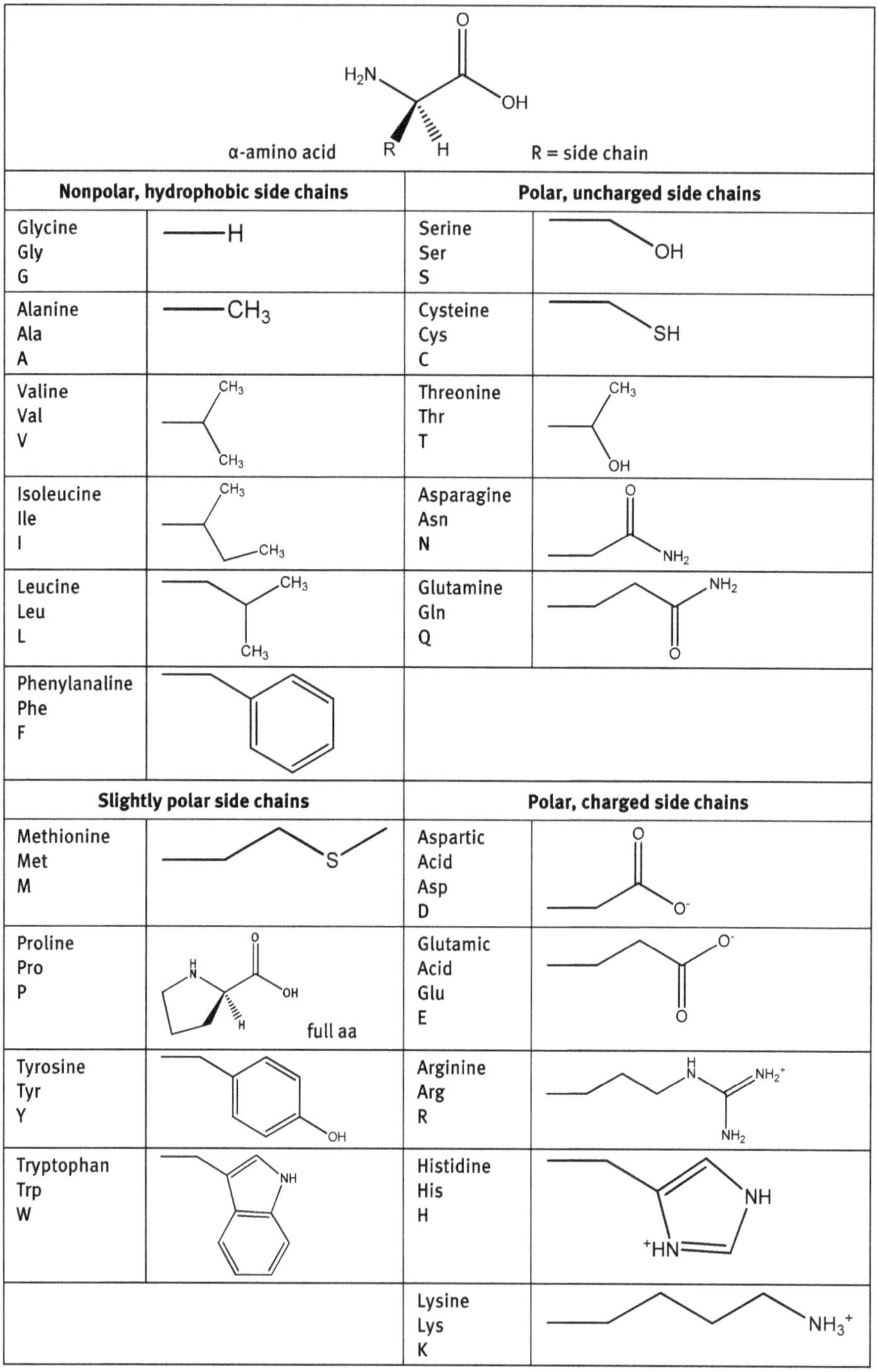

Figure 1.10: The 20 different amino acids that build proteins in humans.

which carries a specific amino acid that is bonded to form a protein with a specific sequence with the help of ribosomes. Since there are 20 different amino acids with different side chains (Figure 1.10), and the length of a protein can be anywhere between tens to thousands of amino acids, the possible range of three-dimensional structures that can be built is large.

The linear chain of amino acids is called the primary structure (Figure 1.11 a). This chain forms either an α-helix, a β-sheet, or a random β-turn (Figure 1.11 b, secondary structure). The α-helix is mechanically elastic and acts like a spring that can stretch as well as bend. The β-sheet can easily stack and form a strong crystal, creating a stiff section in the protein. The β-turn is a flexible linker between the different secondary structures. Additional strength can be gained by disulfide linkages that act as crosslinks. All these secondary structures combined make up the tertiary structure, or three-dimensional structure, of the protein (Figure 1.11 a). If several proteins ag-

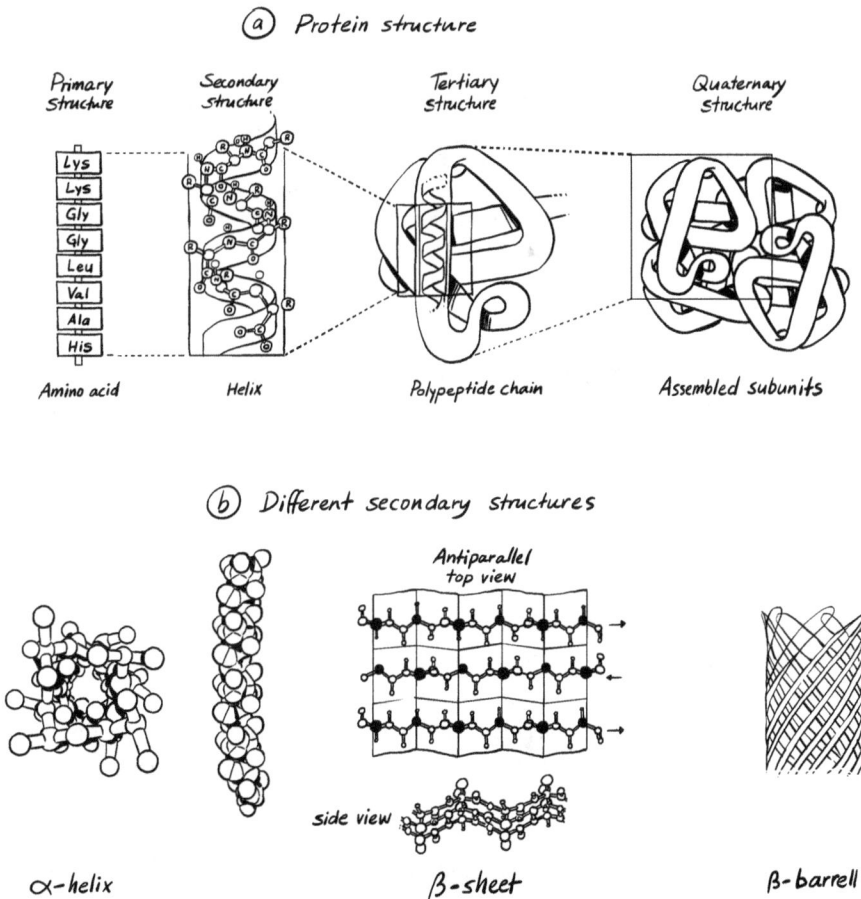

Figure 1.11: (a) Protein structure; (b) Different secondary structures.

gregate to form one, big complex, this is called the quaternary structure of a protein. Hemoglobin is one example, it contains four myoglobins, which aggregate and act together as one large protein complex to transport the oxygen molecules in blood.

Proteins function mainly either as structures or as enzymes. Many of the fibers in cells are made of protein, such as the actin fiber that is part of the cytoskeleton and muscles. Some proteins are active in membrane transport; e. g. a lot of transmembrane ion channels have a barrel-shape made from β-sheets (Figure 1.11 b). Some proteins are antibodies, making use of the specificity of the amino acid side chains to recognize structures from invading bacteria or viruses. A lot of proteins are enzymes, i. e., the catalysts of the cell. These proteins also make use of the specificity to bind a specific metabolite, or substrate, in such a way that the following reaction is lower in energy than it would be unbound (Figure 1.12).

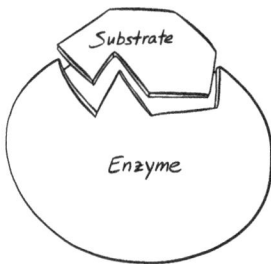

Figure 1.12: The enzyme and its substrate fit together like lock and key. The specificity comes from specific interdisciplinary forces (H-bonding, positive and negative charges) at exactly the right three-dimensional position. The substrate is always positioned so that the bonds that will be broken and the new ones that will be formed are activated.

It is important to mention one specific protein in more detail: the G-protein transmembrane receptors (Figure 1.13) mentioned before in the signal transduction pathways (Figures 1.3 and 1.4). Here, a signal has to be transferred from the outside to the inside of the cell membrane, while at the same time activating an associated enzyme. This is done by lever action. Basically, the transmembrane part of the protein is stiff, so when the signal on the outside binds, that binding is transferred by the force of a lever to the inside, changing the three-dimensional structure of that inside domain. That force moves one of the stiff parts of the associated enzyme and thus creates the active form of the enzyme. This type of lever action is seen in several instances of nanotechnological regulation. Analogous structure-function relationships of ion channels will be discussed in more detail in the neuron section (see Section 1.4).

Structure and function of molecules – monomers and polymers. So far, we discussed the main polymers in the body, or biopolymers. In nanotechnology, artificial, or synthesized, polymers are used as well. Here, the variability of structures is even larger than in biopolymers, since any reactive repeating unit can be used to create a polymer. Common polymers are shown in Figure 1.15. The bonds between the repeating units can simply be carbon-carbon covalent bonds, making the backbone much more hydrophobic in comparison to, for example, the amides of a protein backbone. This is possible since most

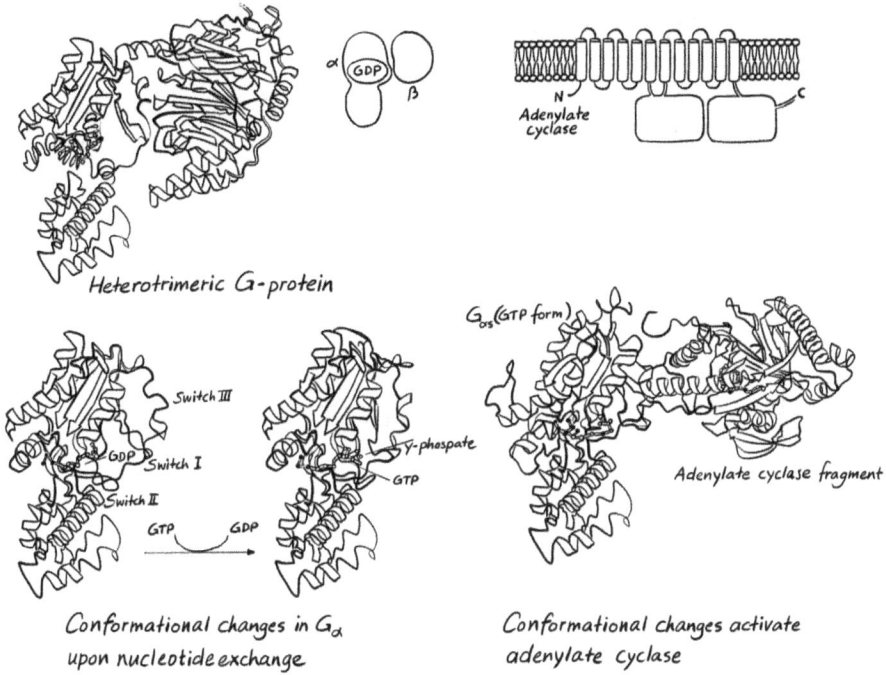

α GDP β

Adenylate
cyclase

N C

Heterotrimeric G-protein

Switch III

G$_{\alpha s}$(GTP form)

GDP Switch I

γ-phospate

Switch II

GTP

GTP GDP

Adenylate cyclase fragment

Conformational changes in G$_{\alpha}$
upon nucleotide exchange

Conformational changes activate
adenylate cyclase

Figure 1.13: G-protein-coupled receptor (GPCR), a transmembrane protein crucial in signal transduction pathways. The transmembrane protein binds a signal at the outside, which triggers a conformational change across the membrane (via lever action of stiff β-sheet sections of the protein) which then activates the enzyme that is attached to the GPCR at the inside of the cell membrane.

Homopolymers

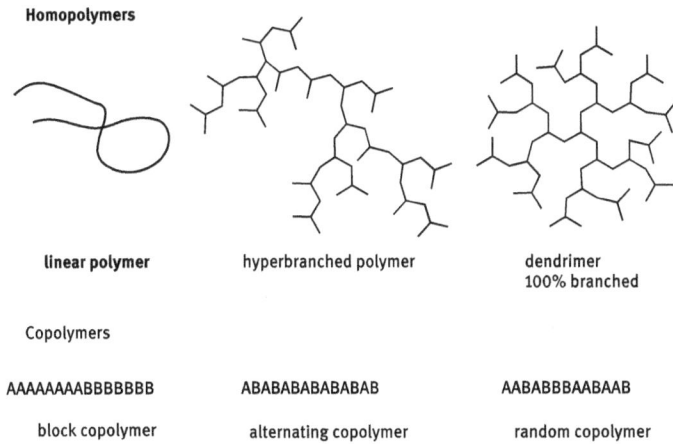

linear polymer

hyperbranched polymer

dendrimer
100% branched

Copolymers

AAAAAAAABBBBBBB

ABABABABABABAB

AABABBBAABAAB

block copolymer

alternating copolymer

random copolymer

Figure 1.14: Possible structures of homopolymers and copolymers.

Name	Repeating unit	Name	Repeating unit
Polyethylene (PE)	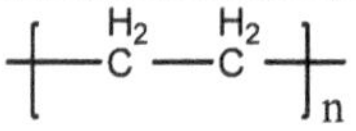	Polypropylene (PP)	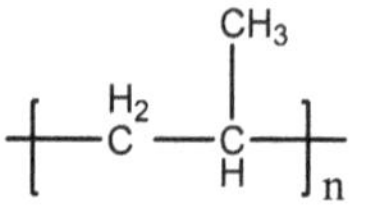
Polybutylene (PB)	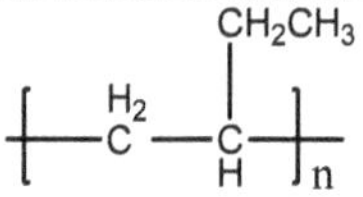	Polyisobutylene (PIB)	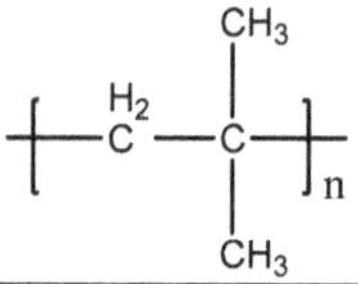
Polybutadiene (PBD)	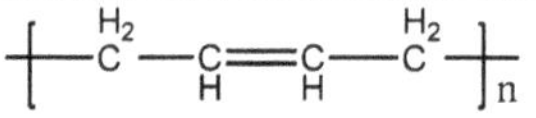	Polyisoprene (natural rubber)	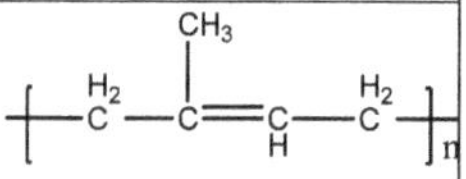
Polychloroprene (Neoprene rubber)	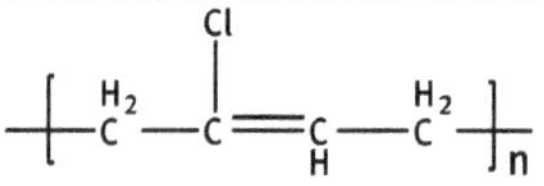	Poly(ethylene oxide) (PEO)	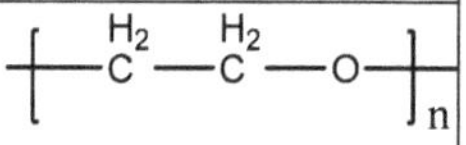
Polystyrene (PS)	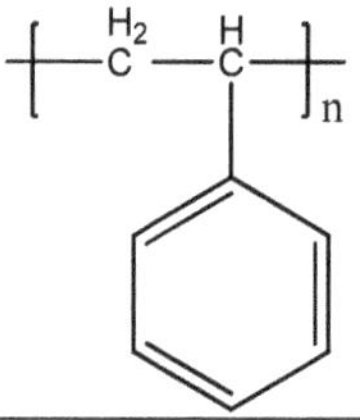	Poly(vinyl acetate) (PVAc)	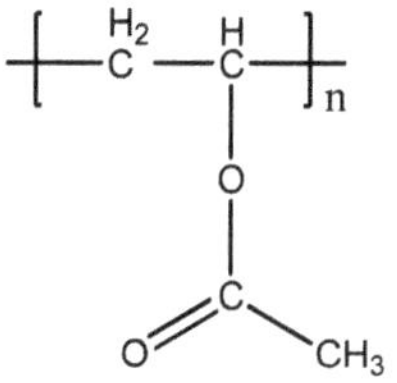
Poly(vinyl chloride) (PVC)	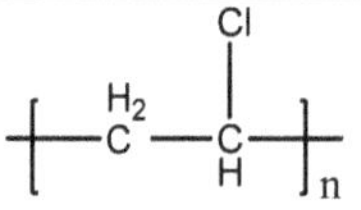	Polytetrefluoroethylene (PTFE)	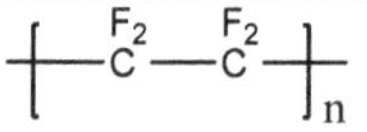
Poly(methyl acrylate) (PMA)	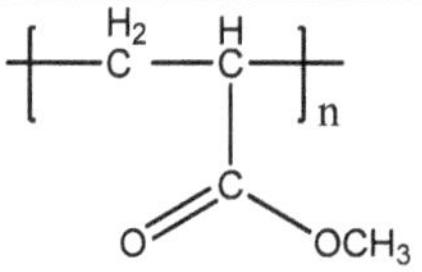	Poly(methyl methacrylate) (PMMA)	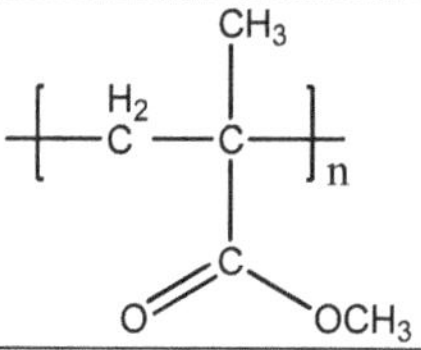
Polyamide (Nylon 6)	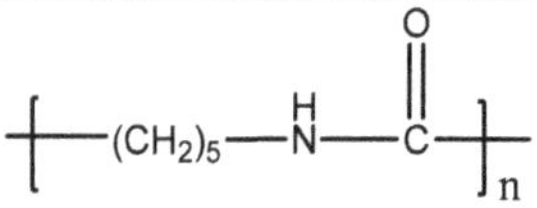	Polycaprolactone (PCL)	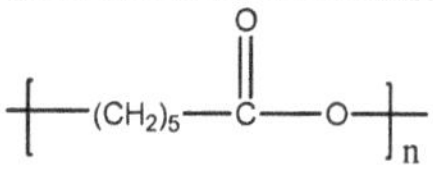
Poly(ethylene terephthalate) (PET)	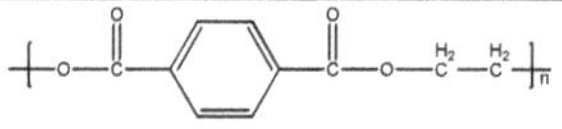	Poly(dimethyl siloxane) (PDMS)	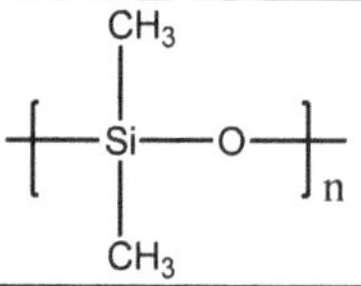

Figure 1.15: Common synthetic polymers and their repeating units.

reactions in the organic chemistry toolkit in any solvent can be used as a polymerization, which would not be possible in the homeostasis-bound cell. On the other end, the cell's enzymes are more specific; in a cell, the length of the polymer is absolutely controlled and there will be no side reactions, which is difficult to replicate in an organic chemistry laboratory.

Synthetic polymers are usually made with just one or a few repeating units (homopolymer and copolymer, respectively). Homopolymers can have different structures: linear of course, but also hyperbranched or dendritic (100 % branched) (Figure 1.14). Copolymers can also be linear, with different repeating units randomly arranged (random copolymer), arranged in an alternating fashion (alternating copolymer), or arranged as blocks of each repeating unit (block copolymer). Alternating copolymers can also be hyperbranched or dendritic. Block-copolymers can be arranged as graft copolymers, where a different block is grafted onto a homopolymer backbone.

Generally speaking, the longer the polymer, the higher its T_g or glass transition temperature and the stronger the material. Strength can be further increased by crosslinking the polymer, as already seen with proteins.

As with protein structures, while sections might be crystalline, the rest of it, the matrix, will be amorphous, i. e., an unordered solid with a T_g but no melting point. Linear polymers will always have a larger T_g than the corresponding branched polymer, and branched structures will not show a melting point (they don't crystallize). The toughest materials are semicrystalline, i. e. contain crystalline regions for strength and an amorphous matrix for elasticity. Also, the stronger the intermolecular forces between the polymers, the stronger the material, i. e., hydrophilic polymers with a lot of hydrogen bonding are generally stronger than hydrophobic ones that only have van-der-Waals forces as intermolecular bonds. But hydrophilic polymers generally dissolve in water unless crosslinked, extremely long, or very crystalline.

A polymer's solubility depends on its hydrophilicity/hydrophobicity. Depending on their structure, polymers can also be amphiphilic and/or charged. If the polymer has conjugated sections, it might be colored (depending on the exact conjugation length). Conjugated polymers can also be conductors or (more likely) semiconductors. Any twist in the backbone reduces the conjugation length, therefore the highest conductivity is found in stiff, crystalline polymers. High crystallinity and molecular weight reduces solubility, thus making it harder to process these materials. The conductivity of a polymer can be further increased by adding conjugated structures such as carbon nanotubes.

Structure and function of molecules – carbon compounds. Carbon molecules like carbon nanotubes are the other set of structures that are common in nanotechnology (Figure 1.16). All of these structures are based on a single sheet of conjoined aromatic rings called graphene. If you stack several sheets together, you form graphite (pencil lead). When rolled up into a tube, it is a carbon nanotube. Shaped into a continuous hollow sphere, it is called a "fullerene", after Buckminster Fuller, who identified the structure. These structures are special in that they are fully conjugated and thus (semi)conductive. Graphene has been used for electrodes in nanotechnology for that reason. Also,

Graphene
one layer of continuous aromatic rings

Graphite
graphene layers stacked on top of each
other to form a layered crystal

Carbon nnaotube
a graphene layer rolled up into a tube

Fullerene
continuous aromatic ring forming a hollow sphere

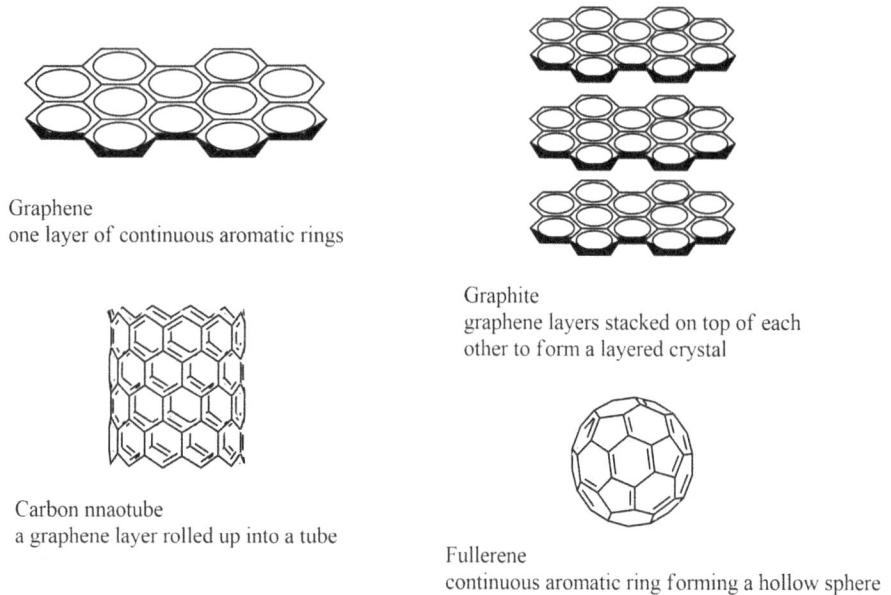

Figure 1.16: Structures of carbon compounds, all based on aromatic rings.

these structures are lightweight but strong considering their weight, thus carbon nan-otubes (as well as larger carbon fibers) have been used as reinforcement or strengthening agents for polymers. Fibers only reinforce materials, though, when they are evenly distributed, rather than phase separated, thus "coupling agents" need to also be added for the strengthening effect. Coupling agents are surfactants, amphiphiles that bind on one side to the fiber, on the other side to the polymer matrix. They reduce the surface energy between the two phases and allow the fibers to mix more easily with the matrix.

Mechanical properties of materials. The strength of the material is an important property. First, let us remind ourselves about mechanical forces. Force is simply mass times acceleration, $F = ma$. The more force I put on an object, the more it will accelerate. The heavier the object, the more force I need to use to accelerate it.

In the case of machines, even on the nanoscale, usually the force has to be transferred somewhere to have an effect. Basically, we are talking about a lever that transmits the force (Figure 1.17). One of the requirements to transfer force, though, is that the lever material is stiff. If the material were flexible, the effect of the force would be to bend the lever; no force would be transferred to the other side of the lever. If the lever were elastic like a spring, it would be even worse: the lever would resist and then force your hand back to where it started (Figure 1.18).

Elasticity is an important property for a material, assuming it is **not** the material for a lever, of course. Elasticity is one of the reasons why some materials don't break so easily, they just deform and then return to their original form when a force acts on them. Brittle materials often break more easily.

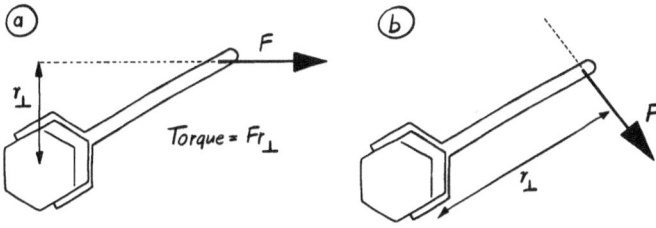

Figure 1.17: a) Torque = $F \cdot r_\perp$. $r_\perp$ is the perpendicular distance from the pivot point to the lever arm, the line of force. Torque is the rotational analogue of force. b) If you make the force perpendicular to the radius, you maximize $r_\perp$ and thus maximize torque. Your force will be most effective in this case. By increasing the length of the lever, your force also becomes more effective.

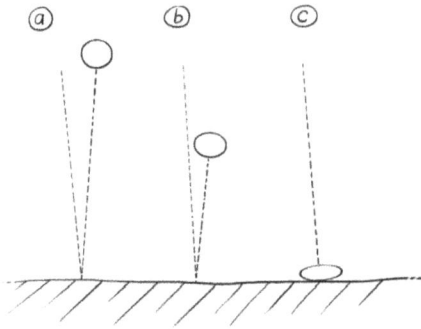

Figure 1.18: Elasticity. a) The ball is fully elastic. The energy of impact was converted into usable energy that carried the ball back up to the height it started from. b) The ball is partially elastic. Some of the energy was usable and it carried the ball back up part of the way, but not to the starting point. The rest of the energy was converted into the deformation of the ball and heat. c) The ball is completely inelastic. All energy of the impact was converted into deformation and heat.

Let us look at the strength of a material in more detail. There are different stresses or forces acting on a material: tensile stress (stretching the material), pressure (pressing on the material – let's start with hydrostatic pressure that equally comes from all sides), and shear stress that comes to the sample from the side (Figure 1.19). In all cases, the force results in strain, which is a change in dimensions proportional to the stress or force. For tensile stress, the sample elongates in one direction, for pressure it becomes shorter in three directions, and for shear stress the top of the sample elongates more than the bottom.

Generally, the strength of the material is characterized by the modulus. This depends on the type of modulus that is measured (it is usually the tensile modulus, or Young's modulus, E). Note that it is strength that we discuss here, not toughness; we will discuss toughness a little later. All of these moduli assume that the material is ideal, i. e., that if the force/strain is removed, the material will go back to its original state instantly. For most materials, this is only true when small stresses are applied. Real materials with

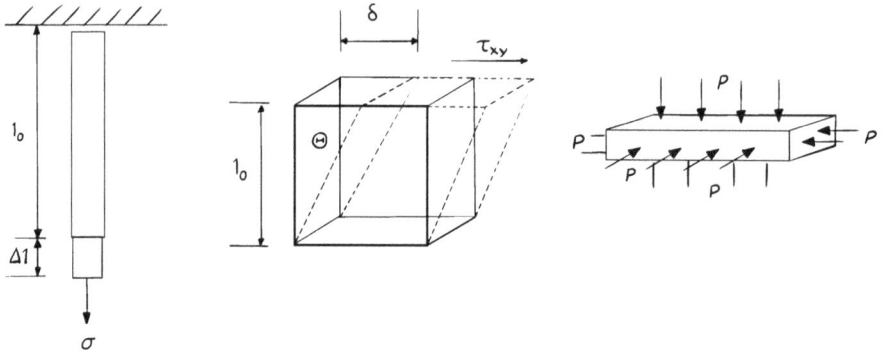

Tensile stress (force) σ:

$$\sigma = \frac{F}{A}$$

Shear stress (force) τ:

$$\tau_{xy} = \frac{F}{A_{xy}}$$

Resulting strain ε:

$$\varepsilon = \frac{l}{l_o}$$

Resulting shear strain γ:

$$\gamma_{xy} = \tan\Theta = \frac{\delta}{l_o}$$

Hooke's law:

$$\sigma = E\varepsilon$$

$$\tau_{xy} = G\gamma_{xy}$$

E: Young's modulus

G: shear modulus

Hydrostatic pressure P

$$P = -B\frac{\Delta V}{V_o}$$

B: Bulk modulus

two-dimensional,
stress and strain on
the same line

three-dimensional
strain is parallel to the
direction of the shear force,
shear force is direction
dependent

three-dimensional

Figure 1.19: A force (strain) always has an effect (strain, elongation) on a material. In the ideal case, stress and strain are linearly related. The conversion factor between stress and strain, the modulus, is a measure of the material's resistance to strain, its strength. The calculations for different forces (tensile or elongation, shear and pressure) are analogous to each other.

larger stresses will permanently deform and/or the effects of stress and strain will take time. The time-dependence of a stress or strain is usually likened to a viscous flow (Figure 1.20).

Yes, even for solids it is considered a flow, just a very slow one. For example, when old window glass sags, it might have taken 100 years to flow, but this movement is still considered a flow. All materials can thus be described as having viscosity. For real materials, you need to combine the effects and equations of strength with the time-dependent effects and equations of viscosity to fully describe their viscoelastic behavior (Figure 1.21). The time-dependent part of the stress is called creep, the time dependent part of the strain is called recovery. In addition, real materials beyond the linear, ideal region of stresses will always end up with some deformation. What is important is that the material can handle the amount of deformation without breaking.

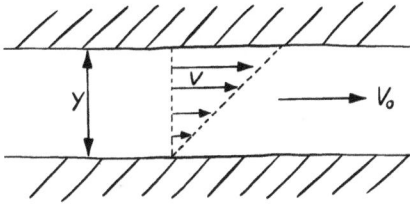

Einstein's law:

$\tau_{xy} = \eta \dot{\gamma}_{xy}$

$\dot{\gamma}_{xy}$: not a strain, but a strain rate

η: viscosity, not modulus

three-dimensional
strain rate is parallel to the
direction of the shear force,
shear force is direction
dependent

Figure 1.20: To more accurately describe non-ideal, time-dependent mechanical strength, modulus is combined with viscosity.

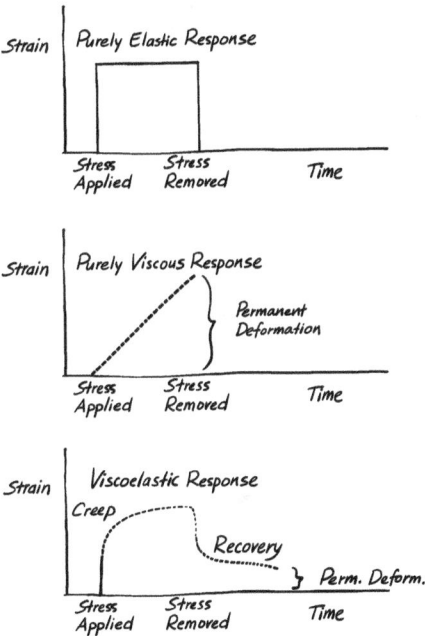

Figure 1.21: With the ideal, purely elastic response all effects are immediate. To describe real behavior, the time dependence of the effect has to be incorporated. This is done by adding viscous flow behavior to the elastic response. The time-dependent application of stress is called creep, the time-dependent removal of stress is called recovery. A large stress past the ideal region also leads to permanent deformation.

And here is where toughness comes in: toughness is the force/volume needed to break the material (Figure 1.22). Basically, you measure the real stress/strain curve until the material breaks and then measure the area under the complete stress/strain curve; that is toughness. Force is the stress, strength is the modulus in the ideal region of a

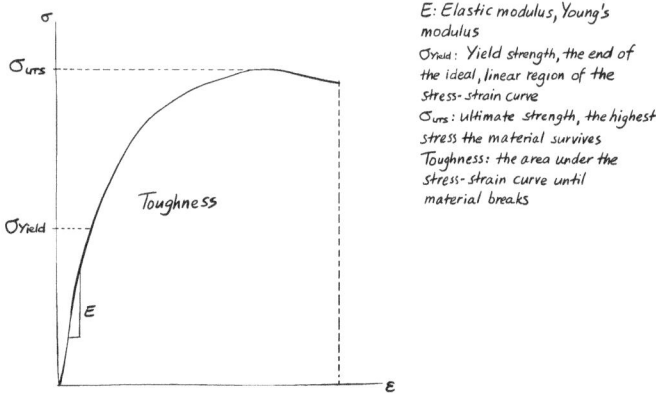

E: Elastic modulus, Young's modulus

σ_{Yield}: Yield strength, the end of the ideal, linear region of the stress-strain curve

σ_{urs}: ultimate strength, the highest stress the material survives

Toughness: the area under the stress-strain curve until material breaks

Figure 1.22: Important measures for the strength of a material.

stress/strain curve, and toughness is the area under the curve of both the ideal and the real region of the curve.

In engineering, it is always important to predict what a material is going to do under ideal and real conditions, and the prediction should be quantitative. So how can one calculate what is going to happen to a material under certain forces in the future? For that, mathematical models are needed. Here, only the most basic ones will be presented.

As viscoelasticity is described as the combination of elasticity and flow viscosity, the models for each will be combined to mathematically express viscoelasticity (Figure 1.23). A spring following Hooke's law represents elasticity, and a dashpot (a liquid-

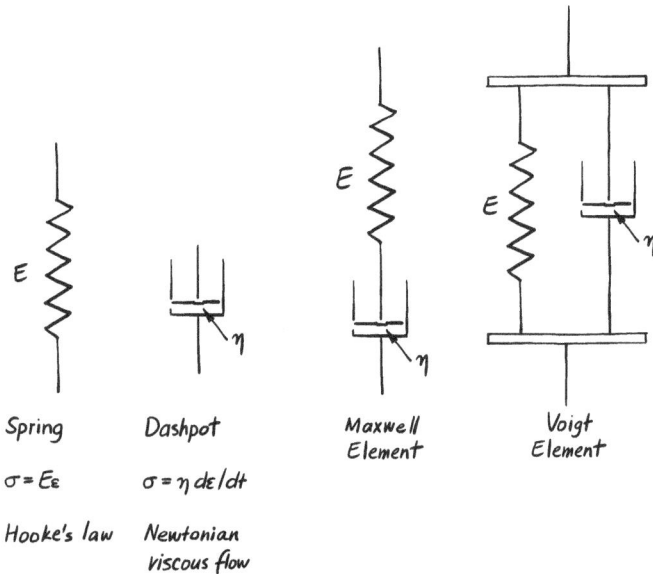

Spring	Dashpot	Maxwell Element	Voigt Element
$\sigma = E\varepsilon$	$\sigma = \eta \, d\varepsilon/dt$		
Hooke's law	Newtonian viscous flow		

Figure 1.23: Basic models of viscoelasticity to predict stress and strain in materials.

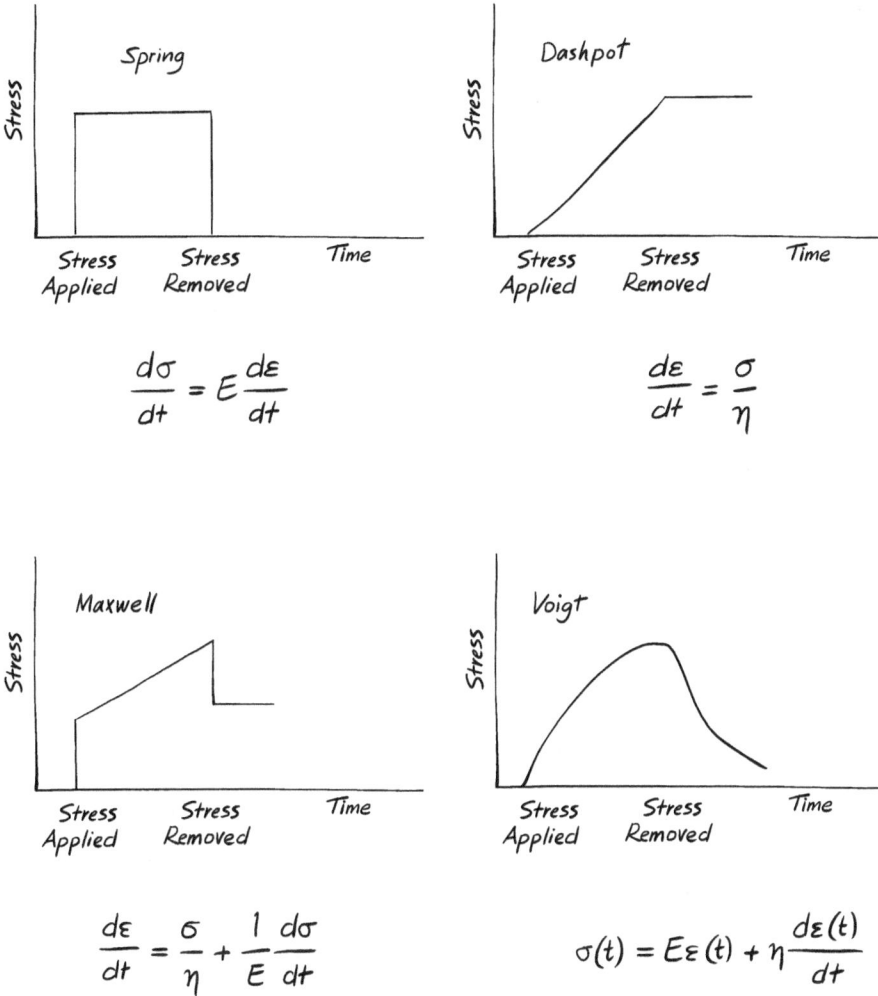

Figure 1.24: Mathematical models of viscoelasticity to predict stress and strain in materials based on springs and dashpots.

Spring

$$\frac{d\sigma}{dt} = E\frac{d\varepsilon}{dt}$$

Dashpot

$$\frac{d\varepsilon}{dt} = \frac{\sigma}{\eta}$$

Maxwell

$$\frac{d\varepsilon}{dt} = \frac{\sigma}{\eta} + \frac{1}{E}\frac{d\sigma}{dt}$$

Voigt

$$\sigma(t) = E\varepsilon(t) + \eta\frac{d\varepsilon(t)}{dt}$$

filled damper) represents viscosity. A Newtonian, i. e. a laminar, ideal flow is assumed. These two models can be combined mathematically in a variety of ways, the most common and simple of which are the Maxwell element (spring and dashpot in series) and the Voigt element (spring and dashpot in parallel) (Figures 1.23 and 1.24). In these models, the equations for stress and strain become mathematical combinations of the equations for ideal elasticity and ideal viscosity.

More complicated combinations of several Maxwell and/or Voigt elements have been proposed. These models are still approximations; stress strain curves of real polymers are more complicated (Figure 1.25). Therefore, in mechanical engineering error

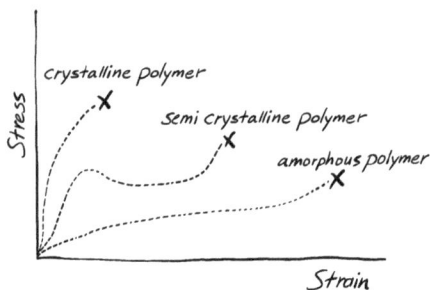

Figure 1.25: Realistic stress-strain curves.

propagations and statistical probabilities of failure are usually built into the mathematical models used.

1.3 Nanoscale Reactions

What are all the chemical reactions these nanoscale actors can perform? Basically, we are talking about all of organic chemistry, inorganic chemistry, and biochemistry combined.... Many of these individual reactions can be found in textbooks written for those fields. Here we will only select a few important facts, principles, and reactions that will help you understand the genius of some of the nanotechnology that will follow.

Let us start with some organic chemistry that will help with biochemistry as well, by looking at the structure-function relationships of organic molecules. Some of the functional groups on an organic or biomolecule are more reactive than others. There are basically two types of structures that are reactive: Double and triple bonds, and polar groups. The reactivity of multiple bonds comes from the type of bond involved: π-bonds. π-bonds are formed by the overlap of p-orbitals. This overlap is energetically a lot less than the overlap between s-orbitals that form σ-bonds. One should watch out for the fact that a double bond consists of a σ-bond and a π-bond, not two π-bonds. Also, σ-bonds can be formed by the overlap of hybridized orbitals, such as the sp^3-hybridized orbitals that from the bonds in CH_4. π-bonds, though, can only be formed by the overlap of p-orbitals. Thus they will always be weaker and more reactive.

The reactive functional groups have in common that they contain polar bonds. Polar bonds are bonds between atoms with significantly different electronegativity or electron density, such as O–H and C=O. The atom with the higher electron density is always the nucleophile, the one with the lower electron density the electrophile. Nucleophiles also have electron pairs that can be used to form bonds, and the electrophile is the receiver of that bond. Hydrogen atoms are generally not considered electrophiles; it is more important to know about hydrogens if a functional group is acidic or basic, i. e., if the proton can leave or get taken up. Polarity and acidity determine partial and full charges; generally, the more charge there is, the more reactive the functional group.

To evaluate the strength of the nucleophile or charge, resonance structures have to be taken into account when applicable; resonance structures distribute charge along several atoms, thus lowering the charge on the specific nucleophile. Another factor that modifies reactivity is steric hindrance, or how crowded a functional group is. A crowded, sterically-hindered functional group is less reactive simply because it is less likely that the other reactant is going to find it.

Another important property of organic and biomolecules is their specific 3-dimensional structure or stereochemistry. When the specificity of binding is mentioned, stereochemistry is a big part of that. For example, the body only produces L-amino acids, D-amino acids are toxic.

There are a variety of stereoisomers (Figure 1.26). Some only change in the rotation of a bond. That is not a different compound, but in instances of e. g. a substrate binding

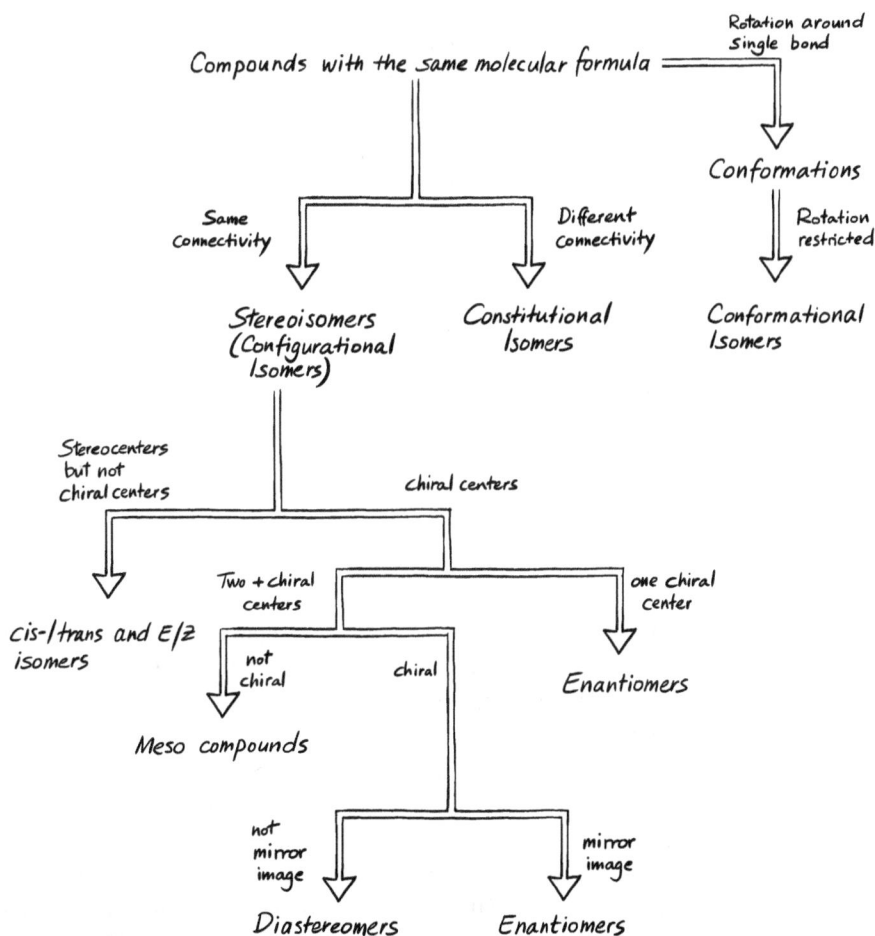

Figure 1.26: Different stereoisomers of organic structures. Cells can distinguish between all of them.

an enzyme, even that is important. Usually, the enzyme ensures that the substrate binds correctly by offering corresponding functional groups creating the correct intermolecular forces, such as a hydrogen donor opposite to a hydrogen acceptor.

Whether double bonds are cis or trans (or E and Z for more highly substituted double bonds) makes a big difference, as can easily be seen in the structure of fatty acids. Cis-isomers are bent and thus have a lower T_g than trans isomers, which are straight chains and crystallize well (Figure 1.6). Using that fact, a membrane can control its stiffness by choosing the correct composition of its lipids.

Stereoisomers that do not change the properties of the molecules but still are crucial in binding and recognition events are compounds with chiral centers, i. e., carbons that contain four different substituents. With one stereocenter or chiral carbon, these two compounds are enantiomers, or mirror images from each other (like our hands, which is why this property is also called "handedness"). The more stereocenters a molecule has, the more compounds there are; if they are not fully mirror images anymore they are either diastereomers or meso-compounds (Figure 1.26).

We looked at the properties and reactivities of organic and biomolecules. What reactions can these molecules perform? There are two broad categories of reactions: radical reactions or polar reactions, i. e., reactions moving one electron versus two electrons, respectively. In Organic Chemistry, radical reactions mostly react with multiple bonds (unless you work with very harsh reaction conditions). With radical reactions, you need an initiating radical (often an initiator molecule split evenly into two), then the reaction propagates itself and only ends when two radicals combine or the monomer runs out (Figure 1.27). The majority of vinyl polymers are prepared in this fashion. In Biochemistry, some oxidation and reduction reactions are also radical reactions with the help of a cofactor or catalyst that can feed or take one electron at a time. CytochromeP450 is such an enzyme. CytP450 oxidizes foreign compounds in the body to make them hydrophilic enough to be excreted in urine (i. e. water soluble). In most cases, these cofactors have metals in their center that have several oxidation states and with that can remove or donate electrons one at a time.

The large majority of Organic Chemistry or Biochemistry reactions, though, are polar reactions where a nucleophile reacts with an electrophile. Organic Chemistry categorizes these reactions by the types of mechanisms (e. g. electrophilic addition to multiple bonds, nucleophilic substitution, carbonyl condensation reaction); biochemistry categorizes reactions by the effect the reaction has on the carbon skeleton of the compound (e. g. isomerization, group transfer reactions, condensation, hydrolysis). Regardless of how the reactions are called, the mechanisms are the same. In Biochemistry, though, most reactions occur via a catalyst, an enzyme. This allows the reaction to take place in water at 37 °C, regardless of what the exact solubility and reactivity is under these conditions. In fact, most biochemical reactions are faster and have less side products than they would in Organic Chemistry because of the enzyme. The other advantage of the enzyme is that it can be precisely regulated, either by binding different substrates differently or not at all, or by binding other factors allosterically (at a different part of the enzyme)

Initiation

Propagation

Termination

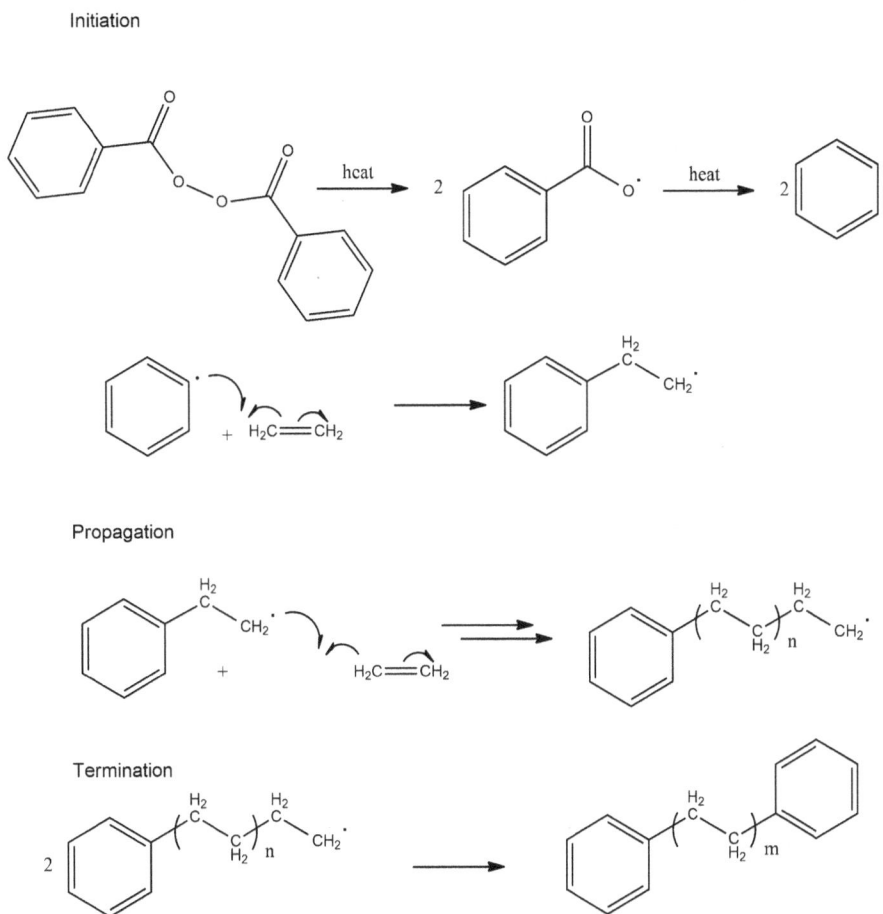

Figure 1.27: An example of a radical reaction (radical polymerization of polyethylene).

and modulating the activity of the enzyme itself. In Organic Chemistry, reactions have to be regulated by environmental factors such as solvent polarity or temperature. And, in almost all cases, the products of Organic Chemistry reactions have to be purified from side products.

One of the most important reactions in both Organic and Biochemistry is the reaction of an acid or a base with water or with each other. The reason for that is first of all that acids and bases change the pH of the aqueous solution. In a cell, this could easily be deadly; remember, cells and organs have to work within the tight confines of homeostasis. Conversely, the pH of the environment changes the reactivity of the molecules in a solution by changing the charge of the molecules, and with that their nucleophilicity or electrophilicity. Since most reactive functional groups are at least somewhat acidic or basic, this interplay has to be taken into account and tightly managed. In the cell that is

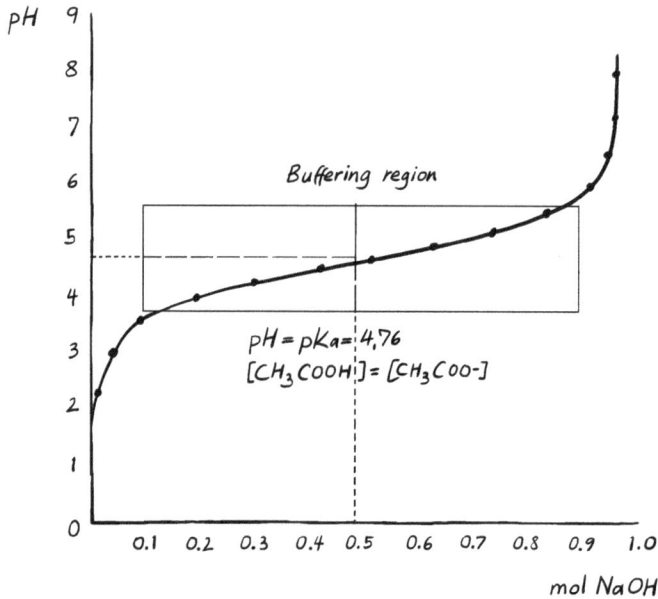

Figure 1.28: Example of a titration curve to determine the buffer capacity of acetic acid.

usually done by having the cell-solution (cytosol) have a certain buffer capacity, so that acids and bases only change the pH slightly when reacting (Figure 1.28).

In Organic Chemistry, when it is necessary to increase the energy of a starting material the reaction is usually heated. This is obviously not possible in a cell. In Biochemistry, the compound is therefore reacted with the "energy-storage molecule" adenosine triphosphate (ATP), and thus a high-energy bond to a phosphate is added to the molecule (Figure 1.29). When this phosphate is then exchanged with a different functional group,

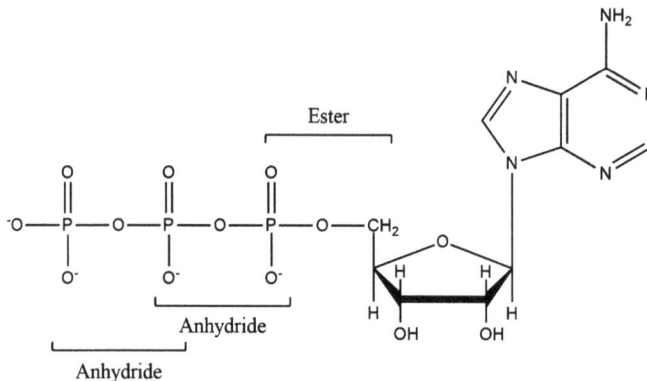

Figure 1.29: Adenosine triphosphate (ATP) is used as the energy storage molecule in the cell.

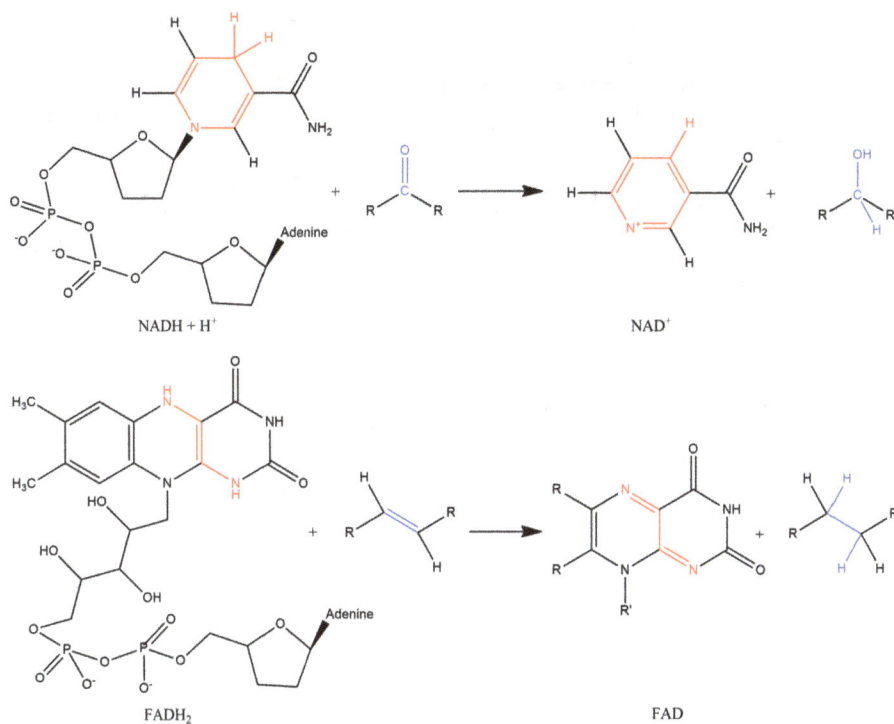

Figure 1.30: NADH+H$^+$ and FADH$_2$ are the reducing agents in the cell.

the reaction is spontaneous, because the product is a lower-energy compound. The other important co-factors in Biochemistry are NADH and FADH$_2$, the reducing equivalents stored in the cell (Figure 1.30). One of the common pathways to generate ATP, NADH, and FADH$_2$ is to go through glycolysis and the citric acid cycle.

Another way to store energy in the cell is to store the molecule that energy is made from. These can be lipids or glycogen. The overview of glycogen and lipid metabolism is shown in Figure 1.31.

Inorganic Chemistry reactions will not be discussed here beyond the oxidation and reduction of metal ions mentioned before. There is, however, an important point to be made about metal nanoparticles: Due to the particles' small size, they consist of a lot of surface and very little bulk. As discussed in Section 1.2, this makes nanoparticles very reactive. Also, when nanoparticles are distributed in a different matrix or phase, due to the large amount of surface area between those two phases this mixture is a high-energy system. The first thing that would happen if left alone is that the system would reduce its energy by aggregating all of the small particles into a large, phase-separated bulk. But since the particles are supposed to stay small, so that they can perform a specific reaction or strengthen the matrix of a polymer, this needs to be avoided. There are a few methods to keep the nanoparticles from aggregating. One is to "pacify" the surface, i. e.

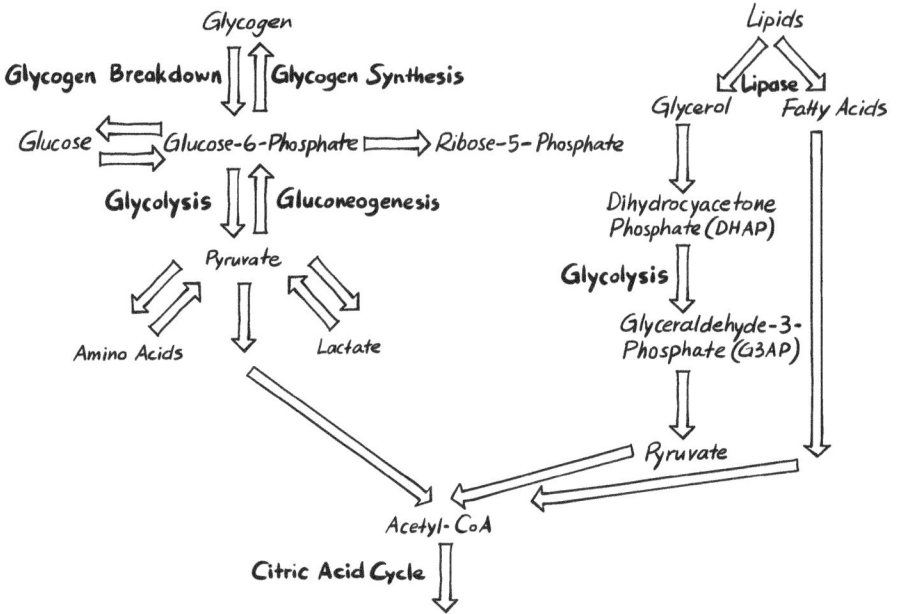

Figure 1.31: Overview of glycogen and lipid metabolism.

surround it with a surfactant that lowers the surface energy. Another is to surround the particle with a surfactant that is charged, so that the particles, all with the same charge on the surface, will repel each other ("charge-stabilization"). One can also simply cover the surface of something bulky, like a polymer that not only reduces the surface energy but also hinders interaction between the particles due to sterics ("steric stabilization"). The same principle can be used for all nanoparticles, not only metal ones.

1.4 The Brain and the Functioning of Nerve Cells

When talking about sensing and movement, one has to also talk about the human nervous system (NS, Figure 1.32). A bit of vocabulary first: "Sensory" (or "afferent") neurons carry information into the central nervous system from sense organs. "Motor" (or "efferent") neurons carry information away from the central nervous system (for muscle control). "Somatic" neurons connect the skin or muscle with the central nervous system. "Visceral" neurons connect the internal organs with the central nervous system.

Automatic responses are carried out by the autonomic peripheral NS (Figure 1.32). The analysis of the senses and controlled, planned movement are performed in the brain. The midbrain controls sensory processes. It relays signals concerned with motor function to other sections in the brain. The cerebrum contains two hemispheres con-

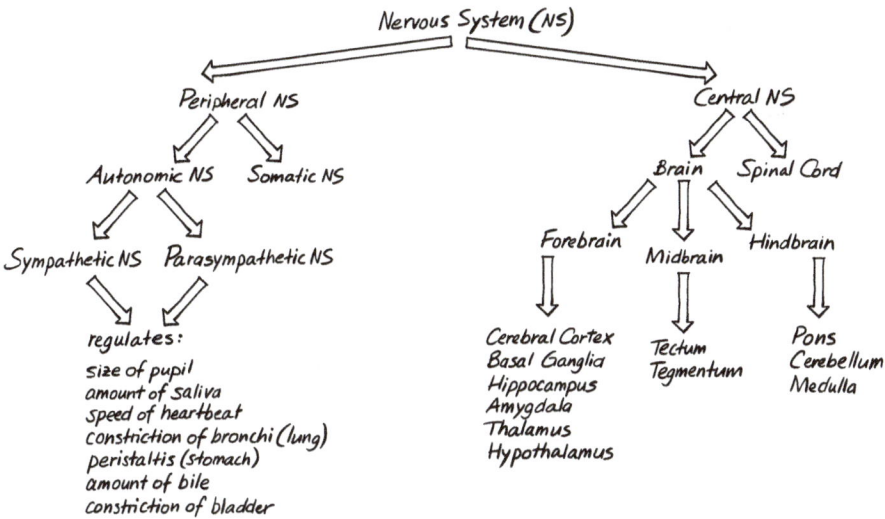

Figure 1.32: The different sections of the human nervous system.

nected by the corpus callosum. The cerebrum is covered by the cerebral cortex. The left hemisphere is connected with the right side of the body and vice versa.

In most people, the left hemisphere controls verbal skills and logic, the right hemisphere spatial perception, art, and music. Each hemisphere is divided into 4 lobes: frontal, parietal, occipital, and temporal.

The cerebral cortex is only 2 mm thick, but is responsible for cognitive functions, such as speech, emotions, memory, and voluntary movement. The cortex is involved in problem solving, emotion, complex thought, coordination of complex movement, initiation of voluntary movement, processing tactile, visual, and sound quality stimuli, processing of multisensory information, and language comprehension and production (Figure 1.33).

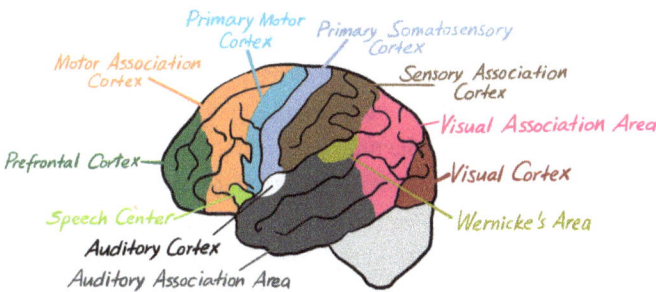

Figure 1.33: The different sections of the cerebral cortex.

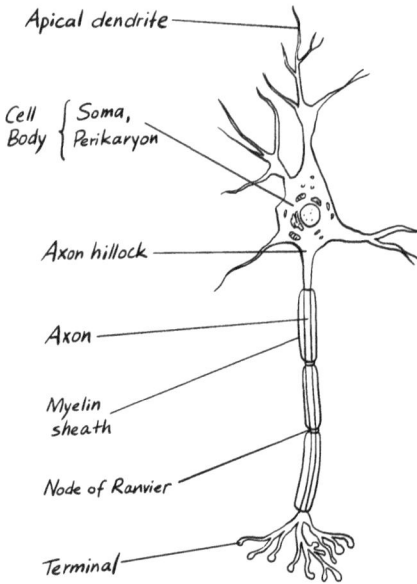

Apical dendrite

Cell Body { Soma, Perikaryon

Axon hillock

Axon

Myelin sheath

Node of Ranvier

Terminal

Figure 1.34: The structure of a neuron.

All of these complex functions are performed with the help of a specialized cell, the neuron. There are several different neurons, but they all have the same basic structure (Figure 1.34). The cell body (soma) has two major extensions: dendrons that receive information from other neurons and axons that deliver information to the following neurons. The incoming information from several neurons will be processed in the soma. If the resulting membrane potential (see below) at the axon hillock is large enough, an action potential (see below as well) will be sent to the following neurons via the axon. The transmission is sped up because of the myelin sheath; basically, the potential can jump from one node to the next.

A "potential" at a membrane is an imbalance of compounds at one side of the membrane in comparison with the other side. This can obviously only happen if the membrane itself is a good barrier and is very selective in what it lets through. This is true for all membranes, but specifically for the membranes of neurons. This imbalance is called "potential" because it is actually a form of energy – potential energy. This "potential" has the potential to perform work, i. e. it is the driving force to balance out the compounds on each side. In the case of the neuron membrane, the" potential" is specifically an electrochemical potential, where not only the number of compounds, but the number of charges is in imbalance. That means that one side has more negative, the other side more positive charges (Figure 1.35). This electrochemical potential at the neuron membrane is generated by having different amounts of specific positive and negative ions on each side, specifically, sodium, potassium, calcium, and chloride ions. To regulate this potential, the neuron membrane contains specific ion channels. Some of them

Ion	OUTSIDE	INSIDE	Membrane potential E_x
K^+	5.5 mM	150 mM	-90 mV
Na^+	150 mM	15 mM	+60 mV
Ca^{2+}	1.5 mM	0.0001 mM	+270 mV
Cl^-	125 mM	9 mM	-70 mV

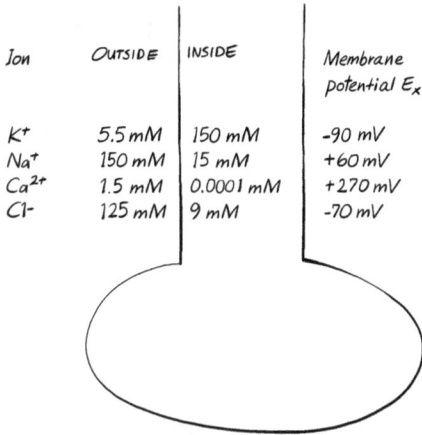

Figure 1.35: The membrane potential of a neuron membrane.

only need to open for a specific ion to equalize. Another one must put in a lot of energy to regenerate the imbalance, the potential.

The sudden, very short equalization of the potential is the "action potential", which is the piece of information sent through the brain (Figure 1.36). Then the original or

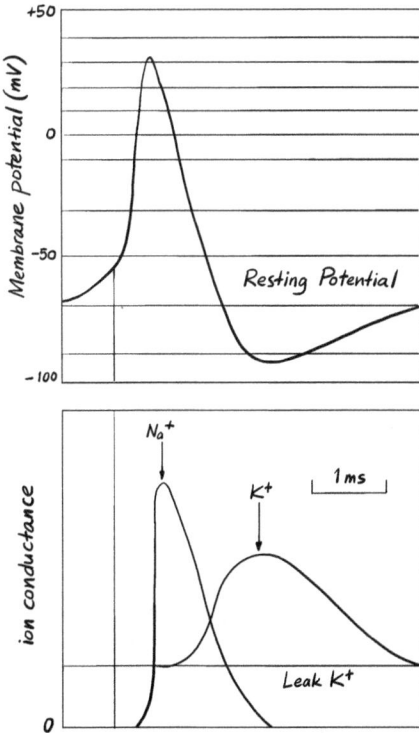

Figure 1.36: The resting potential and the action potential that transmits information along the neuron membrane.

"resting" potential is built up again, so that the neuron can be ready for the next piece of information or action potential. Only two ion channels are part of this process, the sodium channel that releases the potential and the potassium channel that builds it up again. This readiness of each neuron is a high-energy process, which is why the brain uses 70 % of the energy available to the body.

How do these ion channels work? Here is the example of the voltage-gated sodium channel, i. e., the channel that only lets sodium through, but even that only when the membrane potential suddenly changes at the part of the protein that senses voltage (Figure 1.37). Most ion channels have several transmembrane peptide helixes, in this case six helixes together form one subunit, and four subunits together form the channel. The structures in between the subunits fold onto the channel opening; with that the channel is generally closed. One of those sites has a lot of charged amino acids on the outside, which constitute the voltage sensor (Figure 1.37). With a change in voltage comes a change in charge, which changes the three-dimensional structure of the voltage senor. The movement of one section initiates the change in structure of the units that block the channel, thus opening it. The opening of the channel equalizes the sodium concentration on each side, thus starting the action potential.

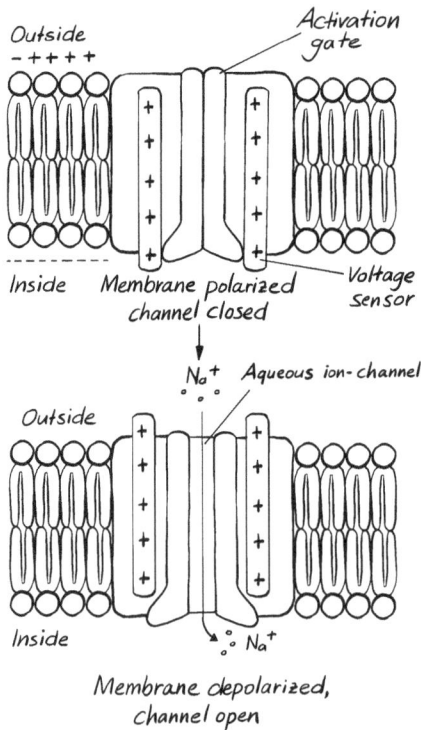

Figure 1.37: How a "voltage-gated" ion channel works. Basically, the channel is normally closed. When the voltage sensor senses a change in the amount of ions, the barriers move away and the channel opens.

There are similar mechanisms for ion-channels not gated by voltage but by other forces. For example, ion channels in the inner ear are opened by having the "plug" pulled away by mechanical forces. And for heat sensing neurons, the ion channel is opened by a structure change initiated by temperature. Other channels open when a specific compound binds (ligand-gated ion channel). In all cases, the opening (or closing) of the channels changes the membrane potential, which either directly creates an action potential in the following neuron, or sometimes first releases a compound that then releases a neurotransmitter, which then generates an action potential in the following neuron.

To fully understand that last paragraph one has to know how the signal is transferred from one neuron to the next, or more specifically, from a neuron's axon to the next neuron's dendrite (Figure 1.38). That end of the axon, the gap between the two neurons called the synaptic cleft, and the beginning of the following neuron's dendrite to-

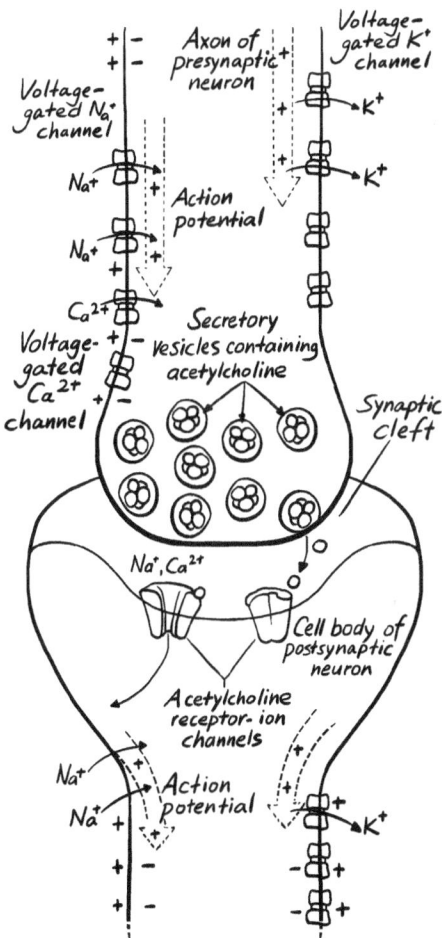

Figure 1.38: How information is transmitted from one neuron to the next by the synapse.

gether are called the "synapse". The membrane of the axon before the synapse has an electrochemical potential, as we have been discussing. That potential is temporarily reversed by an action potential. When that change in potential reaches the tip of the axon, another voltage-gated channel opens and in this case it allows calcium ions to enter. The increase in calcium concentration inside triggers the release of the neurotransmitter acetylcholine into the synaptic cleft via exocytosis (more about exocytosis and neurotransmitter a little later). The acetylcholine is now in the little space between the two neurons, the synaptic cleft. It travels over to the other side of the cleft and binds to a transmembrane receptor – which also happens to be an ion channel. In this case, it is a ligand-gated ion channel and the ligand is acetylcholine. The channel opens and potassium and sodium ions enter the dendrite of that neuron. Since the dendrite also has a potential across its membrane, the influx of all of these positive charges changes its potential and this initiates an action potential that carries the information to the soma of the neuron, where all incoming action potentials are processed. If the processing results in an action potential at the axon hillock, that action potential moves along the axon to the next synapse and the process repeats.

What is exocytosis? Basically, at the axon side of the synapse there are a lot of small vesicles that are held tight by the cytoskeleton (Figure 1.39). They are filled with compounds called neurotransmitters. Neurotransmitters, as their name suggests, transmit information between neurons. There are different neurotransmitters for different types of neurons; also, some of the neurotransmitters will make it more likely and some less likely that the following neuron will end up with an action potential. In the example above the neurotransmitter was acetylcholine, a common neurotransmitter that results

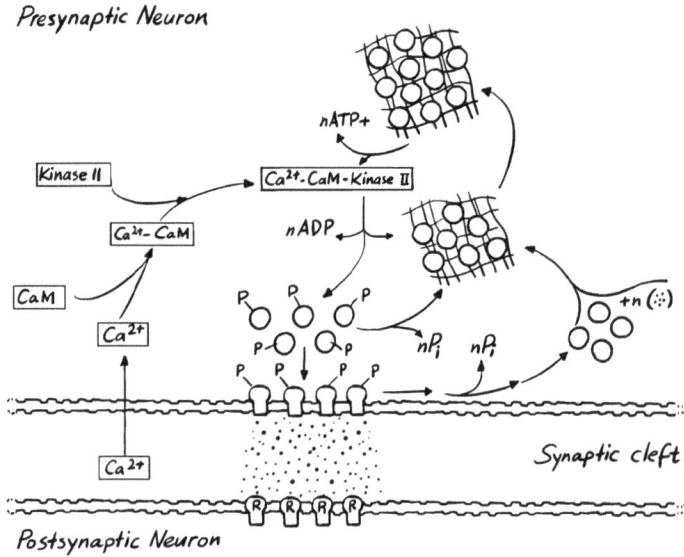

Figure 1.39: Exocytosis into the synaptic cleft.

in muscle contraction. Other neurotransmitters are epinephrine and norepinephrine (also called adrenalin and noradrenalin, respectively), which are both part of the fight-or-flight response, dopamine and serotonin, which are part of mood regulation, as well as histamine, γ-aminobutyric acid (GABA), glycine, glutamate, aspartate, and nitric oxide (NO).

When an action potential arrives at the end of the axon and the calcium channel is opened, the calcium does not only change the membrane potential but the ions themselves also bind to the proteins that hold the neurotransmitter vesicles to the cytoskeleton, thus releasing the vesicles. The vesicles now act like any small particle phase in a matrix (here the solution). The vesicle membranes are hydrophobic, so with the release of the vesicles there is suddenly a high amount of hydrophobic surface in the hydrophilic solution of the nerve ending. To release that high-energy state, the vesicles move to and merge into the synapse membrane. The vesicle and synapse membranes have very similar compositions and can simply merge; this automatically releases their contents, the neurotransmitter, into the synaptic cleft. The neurotransmitter then diffuses to the other side, initiating the action potential in the dendrite, as discussed above.

This is the barebones description of what happens at a synapse. In reality, there are many regulatory mechanisms going on at the same time using other ion channels or transmembrane channels, also signal transduction pathways that modify the activity of the synapse itself, directly or allosterically, all in the name of homeostasis as well as analysis of the incoming data. Here, "analysis" also stands for higher-order thinking, memory, and consciousness. Only a fraction of these details and processes are currently fully understood.

1.5 Machines and Computers on the Microscale and Nanoscale

Machines on the microscale and nanoscale usually end up being based on computer chips. Why is that and how do they work? Let us start with what they made from: semiconductors. In conductors, usually metals, valence electrons have the energy needed to conduct, i. e., are in the conduction band (Figure 1.40).

When the conduction band is energetically somewhat removed from the energy of the valence electrons, you have a semiconductor (Figure 1.40). You can get electrons moving in semiconductors by heating them, to get them energetically into the conduction band. More commonly, voltage is used as energy. To increase the number of charges that are conducted, semiconductors are usually doped.

The most common semiconductor material is silicon, which has a valence of 4. If silicon is doped with atoms with a valence of 3 (e. g., boron) you create "electron holes" and with that a p-type semiconductor. Doping with atoms of a valence of 5 (e. g., phosphorous) will result in an n-type semiconductor that is conducting electrons.

Many parts of electric circuits can be built from semiconductors. An important one is the transistor. A transistor can be both a switch and an amplifier. An example of a

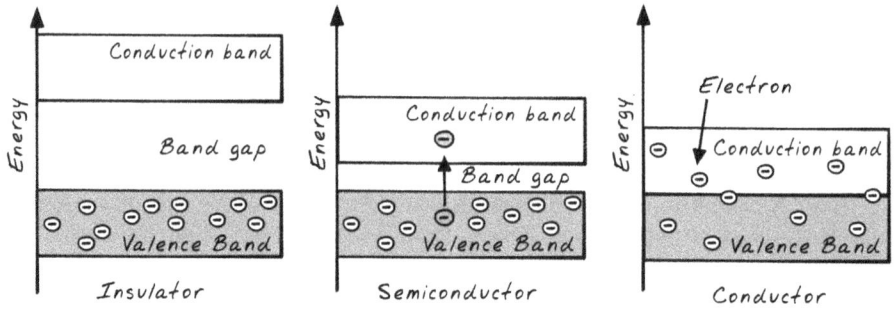

Figure 1.40: Comparison between insulator, semiconductor, and conductor. In a conductor, the energy of the electrons is high enough to be in the conduction band. In a semiconductor, energy has to be added before electrons conduct (i. e., overcome the band gap). In an insulator, the band gap is so high that electrons will not reach the conduction band.

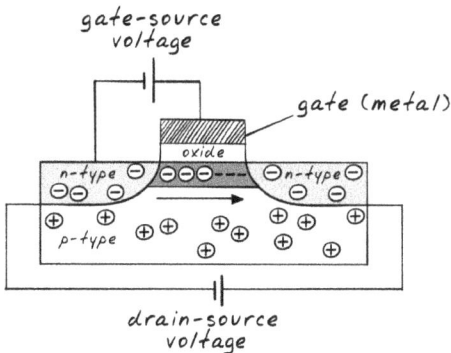

Figure 1.41: An example of a transistor. A transistor combines n-type and p-type semiconductors. The current will only flow when the gate-voltage is high enough to provide more electrons than there are "holes" in the p-type semiconductor.

transistor is the field-effect transistor (Figure 1.41). Basically, a secondary voltage turns the circuit either off or on, which is why they act as switches. The secondary voltage adds enough electrons to neutralize the positive charges of the p-type semiconductor, and when more are added, they create a channel for negative charges, creating the primary voltage. The secondary voltage can also be increased, adding to the primary voltage. At that point, the transistor is not only a switch but also an amplifier. Transistors are often named for the material that they are made of, e. g., MOSFET stands for metal-oxide semiconductor field-effect transistor.

Transistors are often used as switches in logic gates (Figure 1.42). Logic gates add two inputs together (the inputs are always either 0 or 1, as is the output); different logic gates use different Boolean algebra operations. An "And" gate, for example, multiplies the two inputs; the output of that logic gate will always be 0 unless the input consisted of two 1s, in which case it will be 1.

Other important electronic components are resistors that modulate current, capacitors that store charges, and diodes. A resistor lowers the amount of current that passes through (Figure 1.43). Resistance is measured in ohms. According to Ohm's law, the resis-

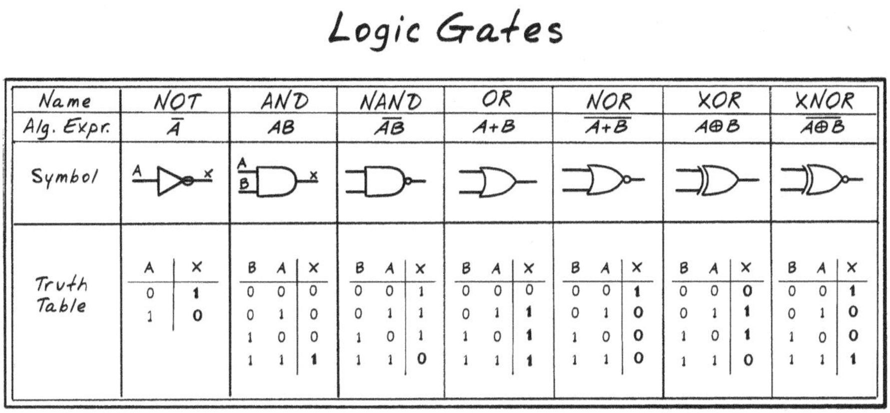

Logic Gates

Name	NOT	AND	NAND	OR	NOR	XOR	XNOR
Alg. Expr.	$\bar{A}$	AB	$\overline{AB}$	$A+B$	$\overline{A+B}$	$A\oplus B$	$\overline{A\oplus B}$
Symbol							

Truth Table

	NOT		AND			NAND			OR			NOR			XOR			XNOR		
	A	X	B	A	X	B	A	X	B	A	X	B	A	X	B	A	X	B	A	X
	0	1	0	0	0	0	0	1	0	0	0	0	0	1	0	0	0	0	0	1
	1	0	0	1	0	0	1	1	0	1	1	0	1	0	0	1	1	0	1	0
			1	0	0	1	0	1	1	0	1	1	0	0	1	0	1	1	0	0
			1	1	1	1	1	0	1	1	1	1	1	0	1	1	0	1	1	1

Figure 1.42: Logic gates, including their symbols and truth table.

tance can be defined as voltage per unit of current. The resistor heats up in the process of hindering the electrical flow; this has led to the resistor's use in heaters as well as light bulbs, in addition to their regulation of current flow.

Diodes are made from a p-n-junction, or the interface between a p-type and an n-type semiconductor, and can transmit electricity only in one direction (Figure 1.44). If the recombination of an electron and hole at the interface results in light emission instead of voltage, the diode is called a "light-emitting diode" or LED.

Capacitors can store charges (Figure 1.45). A capacitor is made up of two metallic plates with a dielectric material in between the plates. A dielectric is an insulator that can be polarized by an electric field to hold charges (air is a common example). When you apply a voltage over the two plates, an electric field is created. A positive charge will collect on one plate and a negative charge on the other, thus separating the charges.

If you integrate several components into one discrete circuit it is called an "integrated circuit" (IC), or "(micro-)chip". If the system involves a micro-sized or nano-sized moving part, the chips are called micro-electrical-mechanical systems (MEMS) or nano-electrical-mechanical systems (NEMS), respectively. One common moving, or mechanical, device often incorporated in a circuit is a piezo-electric material, i. e., a material that generates a current when mechanically stressed.

Besides LEDs, there are also phototransistors, photodiodes, and photoresistors. All work the same way as their electronic counterparts, but they are initiated by light instead of voltage. All can be incorporated into ICs.

Other possible components of ICs are optoelectronic devices, display technologies, magnetic (inductive) devices, or, if you are talking about sensors, sensing or detection elements, transducers, and detectors. In other words, an IC could contain a complete sensor. We are going to talk about sensing or detection elements, transducers, and detectors further below (Section 1.7).

a) Resistor classification

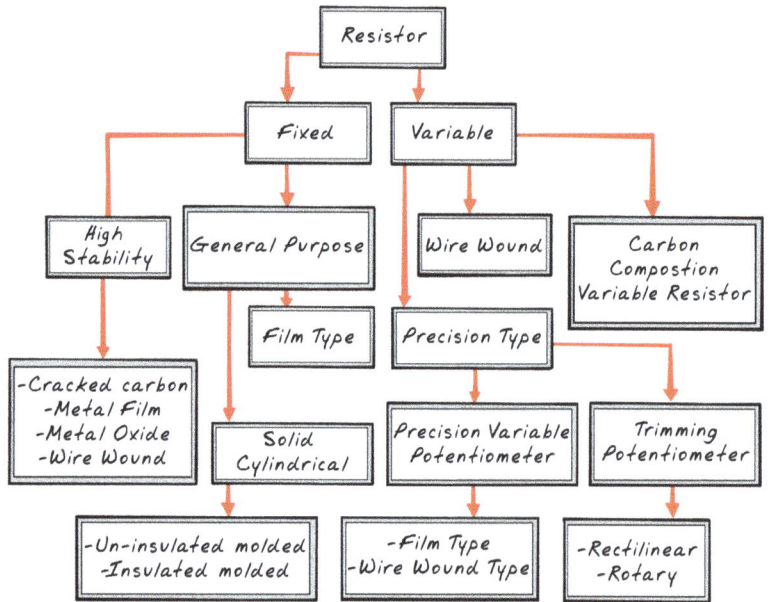

b) Carbon film resistor

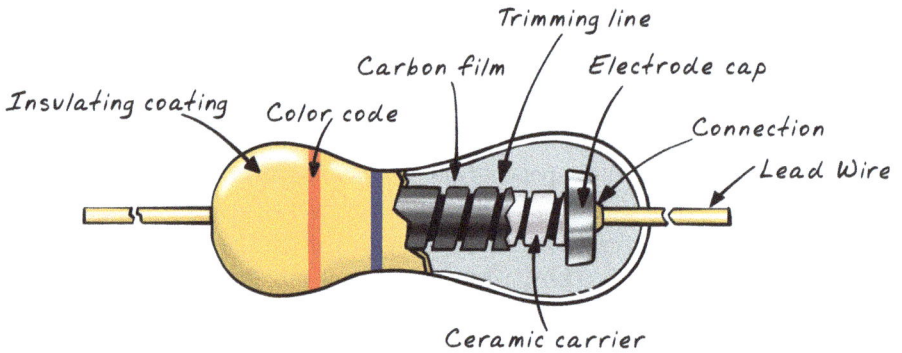

Figure 1.43: Resisters control the amount of current flowing through a circuit. a) Classification of resistor types; b) Components of a resistor (here: carbon film resistor).

When talking about energy storage, batteries are important (Figure 1.46). They are made from electrodes (anode and cathode) in contact with an electrolyte, together called an "electrochemical cell". In the battery, an oxidation reaction occurs that generates electrons and thus a current. In rechargeable cells, the oxidation is reversible. A commonly-

(a)

(b)

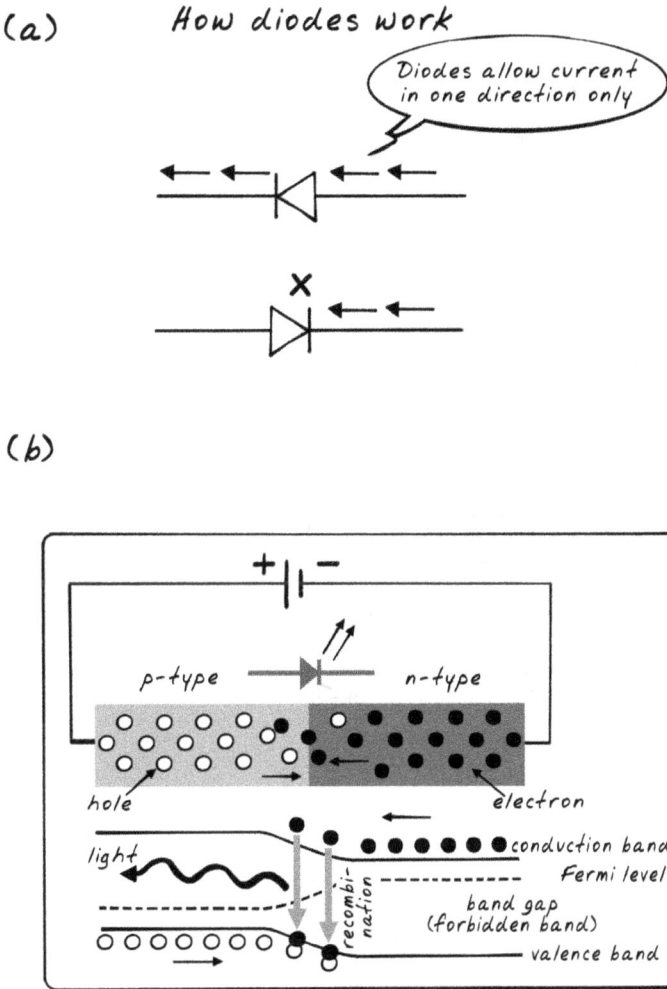

Figure 1.44: Diodes are "PN Junctions", i. e., the interface between a p-type and an n-type semiconductor. a) Diodes allow the current only in one direction; b) In light-emitting diodes (LED) the recombination of the electron with the hole results in light emission.

used electrode is the indium tin oxide (ITO) electrode, since it is very conductive yet transparent.

Nowadays, chemical reactions can also be performed on or in chips (lab-on-a-chip). The technology to create small channels in silicon exists due to chip manufacturing. Therefore, it is possible to flow one starting material in one channel, another in a second channel, letting those two channels meet and the starting materials react, and then flow out the product. This can also be combined with a sensor, e. g., an IR detector that determines if the reaction was successful. There is one caveat that should be mentioned:

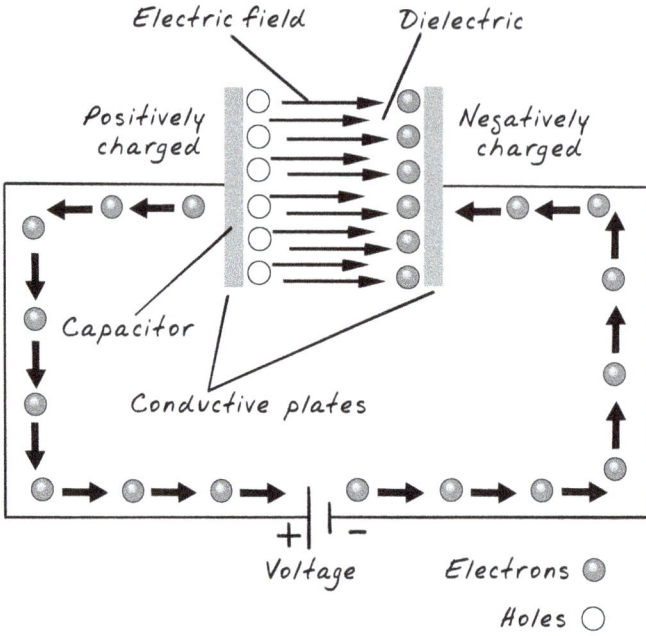

Figure 1.45: Capacitors are two conductive plates with a dielectric (insulator) between them. The plates can be holding charges, one plate for positive and the other for negative charges, thus separating the charges.

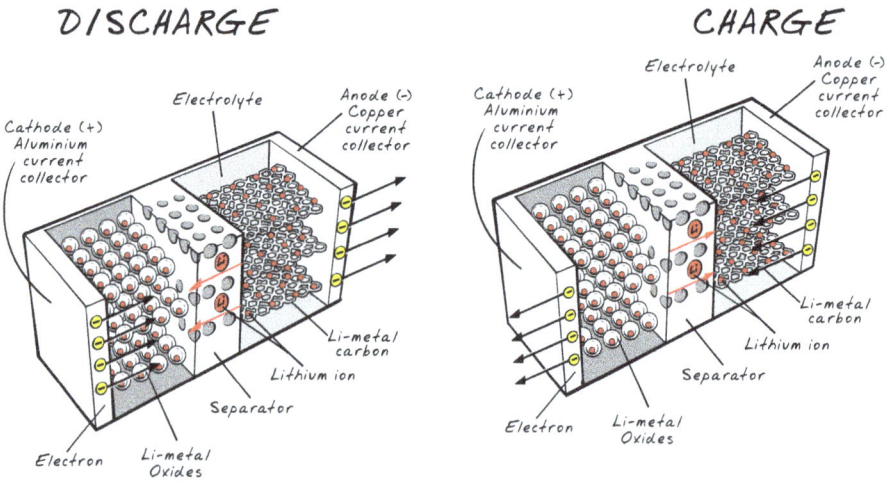

Figure 1.46: Batteries perform chemical reactions at the electrodes to generate electricity. With rechargeable batteries, the reactions at the electrodes are reversible.

on this small scale, as always, surface effects predominate. That could lead to unintended consequences, e. g., a starting material reacts with the silicon channel surface, the roughness of the surface creates turbulent flow, hindering the reaction, or the resulting viscosity stops all flow.

The analysis of any sensor signal is important; you must not only detect a signal, but also know what the signal means. Looking at vision, for example, the only things eyes detect are a large number of photons of different energy. Only after analysis takes place in the brain, will you know that these photons mean that a lover is handing you a red rose. That brain analysis is also the basis of intelligence and consciousness.

"Artificial Intelligence" (AI) is the attempt to mimic intelligence with computers. It started out by optimizing the logic operations and algorithms to make each of the operations faster. This allows for quickly comparing different options and then choosing the "best" option based on a set of parameters. Another AI method works with networks, as the brain works with neural networks. A set of inputs acts on a lot of "neurons", that layer of operations then operates on the output layer of the network. These networks can be optimized by "learning". One common and established way to "machine learn" is by pattern recognition (Figure 1.47). The computer basically "memorizes" the most common output patterns, and that information is fed into the intermediate network layer via a feedback loop. The more operations that are performed, the more common outputs identified, and thus the output becomes more accurate. This is how the computer Deep

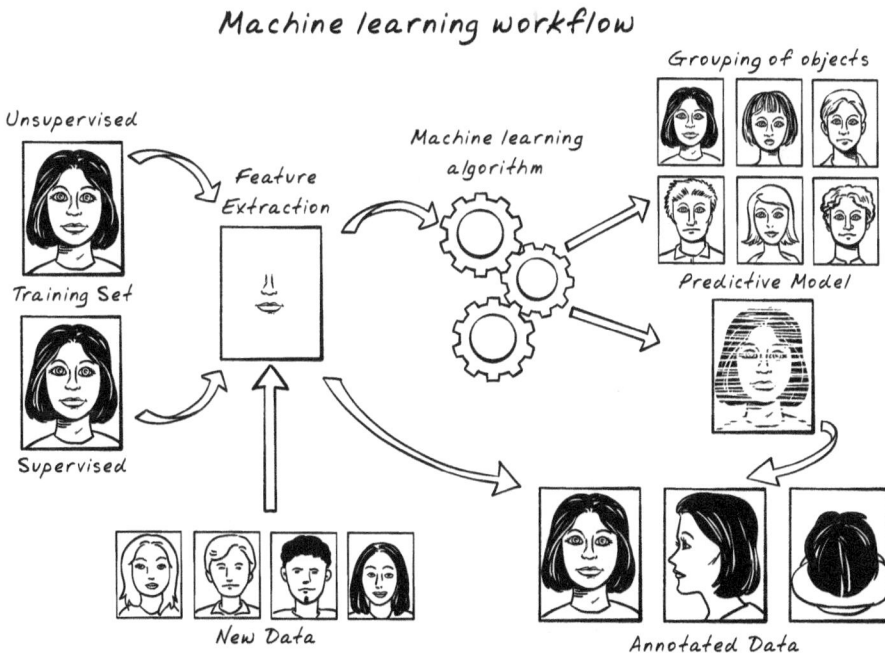

Figure 1.47: Machine learning workflow.

Blue defeated the grand master Garry Kasparov in chess. More effective techniques of computer learning coupled with smaller and smaller transistors, and thus faster and faster processing speeds, have resulted in constant improvement and are impressive feats of artificial intelligence.

1.6 Detection Methods

Before a signal can be analyzed, it has to be detected. But before a compound can be detected it often has to be isolated or purified. In Organic Chemistry, one of the most common separation techniques is column chromatography (Figure 1.48). Column chromatography uses a solid and a liquid phase. The sample mixture is dissolved in the liquid phase, then moved through the solid phase. Interacting with the solid and the liquid molecules, the different compounds are held back differently and thus are separated. That interaction can be based on different polarity, charge, or size. High-pressure liquid chromatography (HPLC) is a column-chromatography that is based on polarity differences, gel-permeation chromatography (GPC, also called size-exclusion-chromatography, SEC) is based on size differences (Figure 1.49), and ion-exchange chromatography (IEC) is based on charge differences. Sometimes, special solid phases are prepared that are molecularly imprinted. Basically, the polymer for the solid phase is

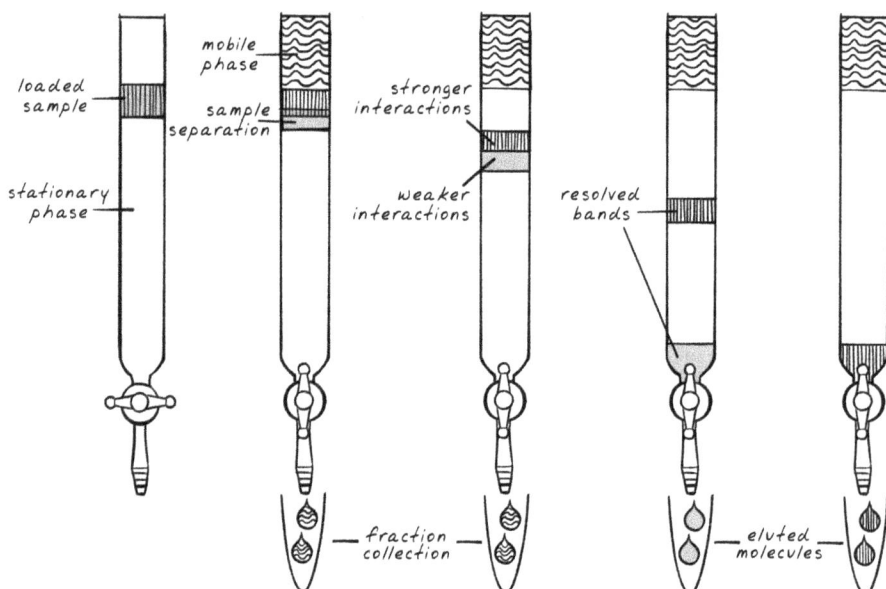

Figure 1.48: Column chromatography. The different components of the loaded sample adhere differently to the solid and the liquid phase. Compounds with good solubility in the liquid phase and poor adhesion to the solid phase will elute first.

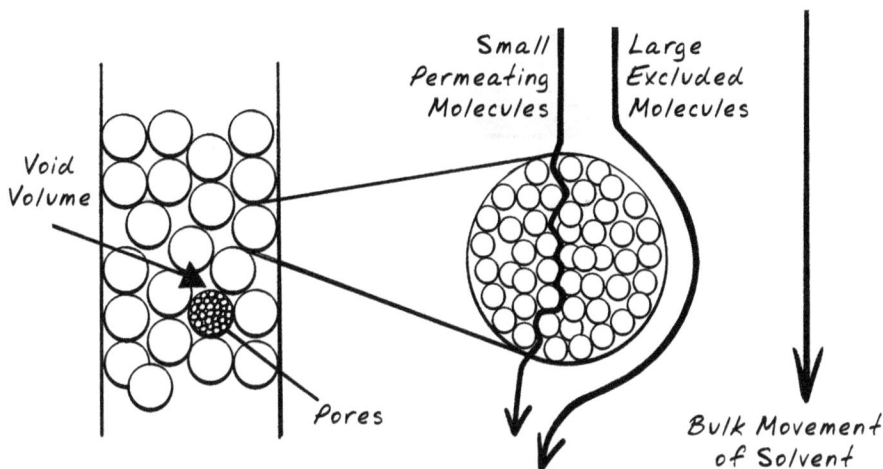

Figure 1.49: Gel permeation chromatography (GPC), also called size-exclusion chromatography (SEC). The solid phase of the column contains pores of a specific size. Molecules of a size smaller than the pores are often retained, and thus are eluted last.

synthesized in the presence of a small molecule template. The template is then washed out, but the imprint specific to the template stays behind. This method was originally designed to separate out enantiomers but now has also been developed for enzyme mimics and other applications. In all chromatography, the different fractions are collected and then analyzed by a variety of methods and detectors.

There are a large variety of detectors. Most of them are based on weight, light, pressure, or electricity output. For nanotechnology applications, the most common detector based on weight is the quartz crystal microbalance (QCM) (Figure 1.50). It does not really measure the weight of the crystal but its resonance frequency, and the resonance frequency is dependent on the weight of the sample. So, if something binds to the surface of the crystal, a shift in resonance frequency is detected.

There are a variety of detectors based on light (Figure 1.51a). Light absorbance can be measured with an UV/Vis spectrometer, light emission by a fluorimeter. Absorbance is based on the composition and concentration of a sample. Therefore, the absorbance can also be used to measure concentration using Beer's law. Fluorescent molecules, when excited with a specific wavelength, emit light in a different and higher wavelength (Figure 1.51b). Fluorescence is used in a variety of ways. There are fluorescent dyes that change their emission spectrum when they are in different phases. That can be used to measure the formation of a specific lipid phase. Fluorescence is often used to detect the location and concentration of a metabolite in a cell by using fluorescent antibodies (Figure 1.51c). Luminescence can be used similar to fluorescence. Specific fluorescent or luminescent dyes are used; dyes for a large variety of experiments have been developed.

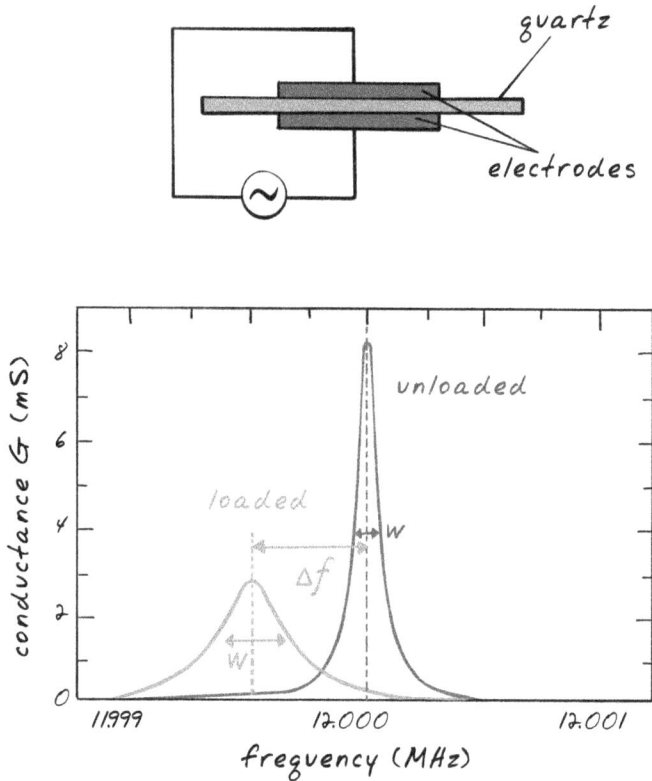

Figure 1.50: The quartz crystal microbalance (QCM) consists of a quartz crystal sandwiched between two electrodes. What is measured is the resonance frequency of the crystal, which changes depending on the weight put upon the crystal.

The most common method to detect pressure on the micro- and nano-scale is piezoelectricity. As already explained in Section 1.5, a piezo-electric material generates a current when mechanically stressed. There are different piezo-electrical materials, and each material can be calibrated, resulting in very sensitive pressure measurements for a wide variety of pressures.

The most common output for sensors is a specific electrical voltage or current, or a charge on a surface. Potentiometers of all sizes can be used to measure voltages. Static charges can be measured by coulombmeters (Figure 1.52). Measuring the current of a living cell, a cell membrane, or an ion channel is somewhat more complicated. For that, the patch-clamp technique is used (Figure 1.53). Basically, an electrode in the form of a pipet is attached to a membrane section, and the current of that electrode is compared with the current of a reference electrode in the surrounding solution. There are some electrodes that can also be used to measure currents inside a cell.

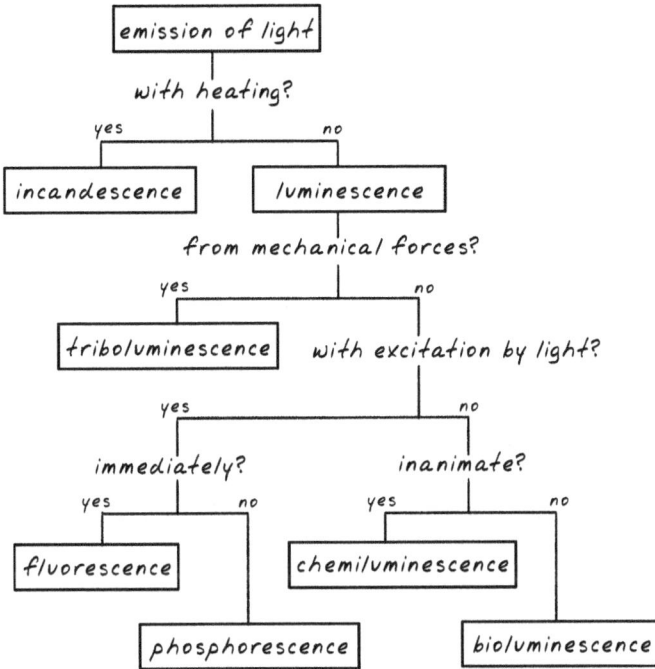

Figure 1.51a: These are all possible types of light emission. All are used in analytical measurements for different applications.

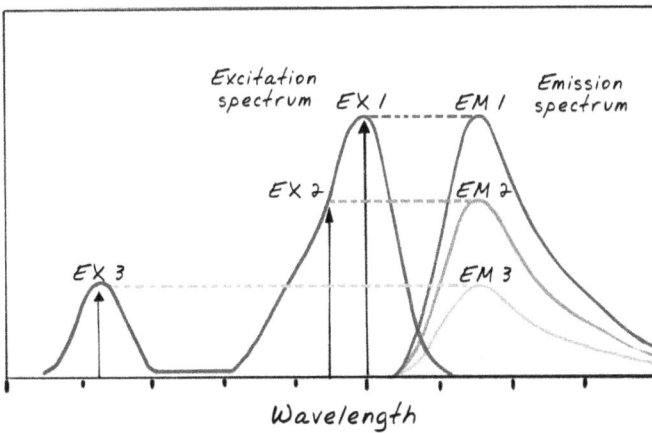

Figure 1.51b: Fluorescent molecules when excited with a specific wavelength emit light in a higher wavelength.

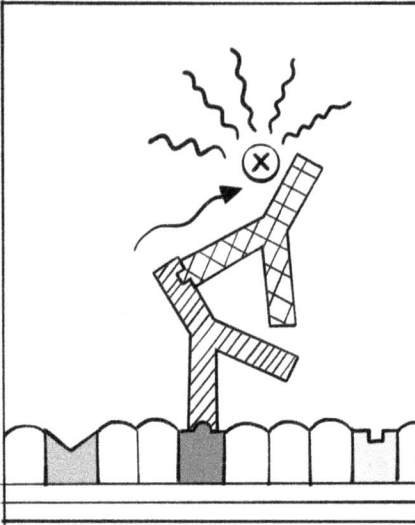

Figure 1.51c: A fluorescent antibody is emitting when biding to another antibody, which is bound to a specific protein. This can be used to determine the concentration of a specific protein in a mixture or even in a cell.

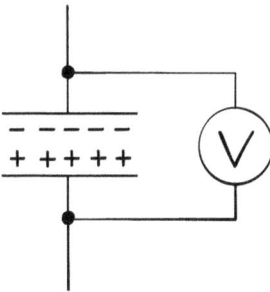

Figure 1.52: A coulomb meter measures the voltage across a capacitor.

Another electrophysiological method mentioned in this book is the electroantennogram (Figure 1.54). Basically, the difference of voltage between the two ends of an insect antenna is measured. Antenna are the noses of insects, therefore, electroantennograms are used to determine how the olfactory sense works in insects and what insects can smell. It has been used to detect a variety of chemicals.

1.7 Sensor Elements and Testing on the Nanoscale

This is a long introduction, and we haven't really talked about what it is all about yet! The book is about human senses and movements, how they work in the nanoscale, and how they have been mimicked on the nanoscale. Therefore, we need to talk a little more about sensors in general before we can start describing current knowledge and approaches.

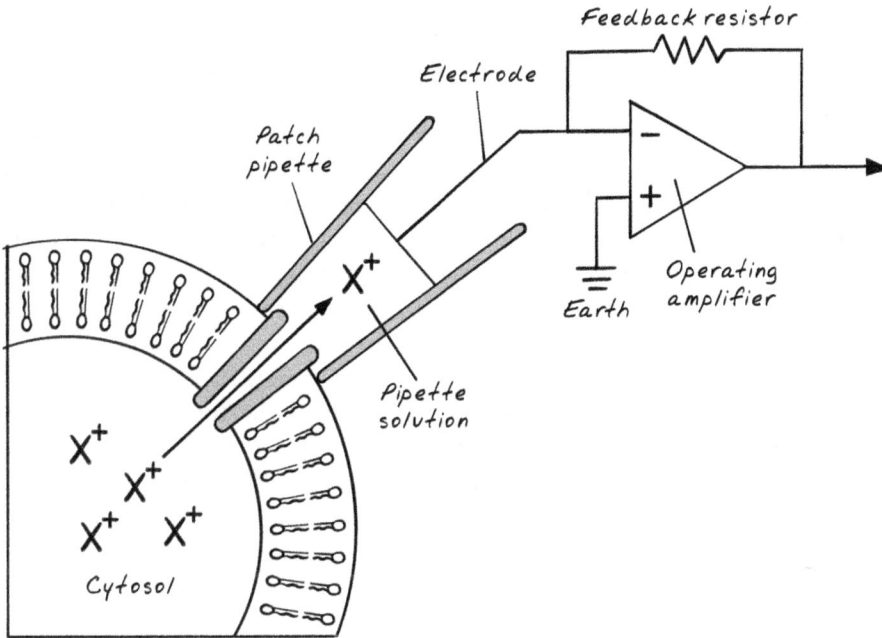

Figure 1.53: The patch clamp technique is used to measure the current of a living cell, a cell membrane, or an ion channel in a cell membrane.

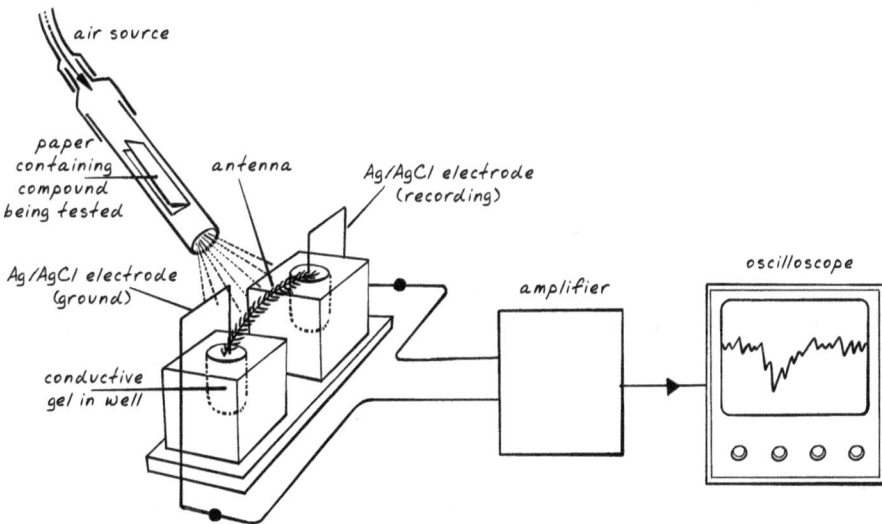

Figure 1.54: Measuring an electroantennogram. The voltage activity of an insect antenna will be measured by connecting each end of the antenna to a different electrode.

Sensors always have several elements: the sensing or detection element, the transducer, possibly an amplifier, and an analysis element. What does that look like?

In humans, the sensing elements, or detectors, are the eyes, nose, ears, and the mechanical and temperature sensors in the skin. The transducers are often receptors or ion channels that change the original signal into an electrical signal, more specifically a membrane potential. That membrane potential is or initiates an action potential that is sent to various places in the brain. In the brain, the signal will be analyzed and made sense of. The brain also initiates actions based on the sensor input, i. e., your hand might take the rose your lover hands you. And it will often have some memory of the signal and can learn from it.

In technology, sensors work very much in the same way. The signal is generally transduced into another signal that can be analyzed more easily. The fuel gauge in your car is a good example (Figure 1.55): the fluid level is hard to measure, but when transduced to an electrical signal via a resistor it is easy to analyze. The most common amplifier in (nano)technology is the bipolar junction transistor. An additional voltage is adding to the charges of the transistor output, and thus amplifies the signal.

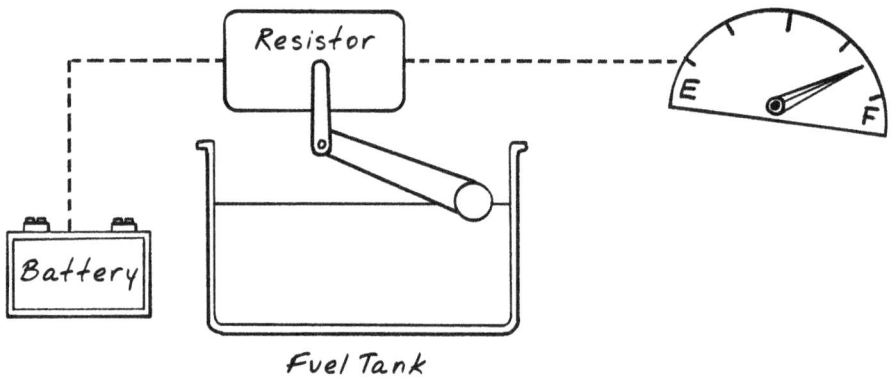

Figure 1.55: Example of a transducer: the fuel level is sensed, then transduced into an electrical signal via a resistor, which is then analyzed and displayed on the dashboard of your car.

In biomimetic nanotechnology, there are a variety of sensing elements, but in a lot of cases the original signal is translated into an electrical signal that is transferred into a computer for analysis and recognition. Nowadays, computers might even remember the signal or a signal pattern and learn from that "experience".

When developing and creating a new sensor in the nanoscale, one always has to prove that one actually made what one wanted to make as well as calibrate the sensor. Therefore, the new sensors always have to be characterized. There are only few techniques that can characterize materials on such a small scale. For conductive and semiconductive materials, the most common technique is the scanning tunneling microscope (STM) (Figure 1.56). When two conductive surfaces are getting very close to each other,

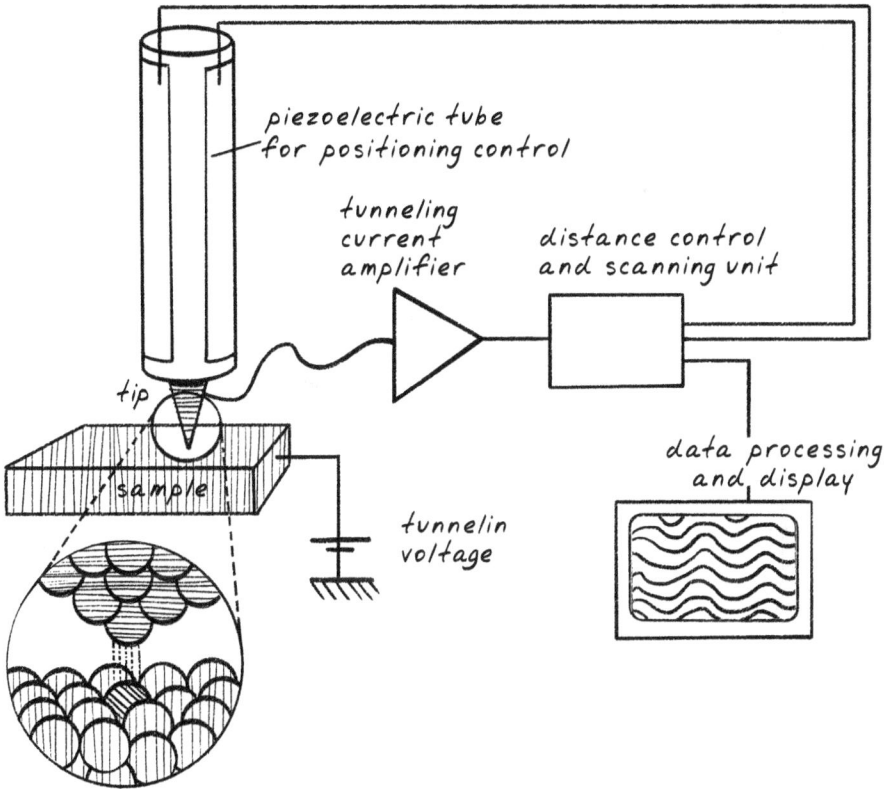

Figure 1.56: In a scanning tunneling microscope (STM), the tiny tunneling voltage between the atomically sharp tip and the surface is measured and amplified. With a precise control over the distance between the tip and the surface, this can be used to characterize the topography of the sample.

a tiny voltage between the two surfaces develops (the "tunneling voltage"). This voltage can be amplified and measured. The STM keeps the tunneling voltage constant while measuring the height of the tip very accurately and with that can scan the surface and characterize the topography of the sample.

A similar method can be used for nonconductive surfaces: An atomic force microscope (AFM) measures the repulsive force between two surfaces by measuring the movement of the tip with a laser (Figure 1.57). The AFM keeps the repulsive force constant, while measuring the height of the tip very accurately and with that can scan the surface and characterize the topography of the sample. But since the measurement is based on a force, AFM can also be used for force measurements on the nanoscale (Figure 1.58). Depending on in which direction the force of the tip is applied, friction (via intermolecular forces on the surface) or elasticity can be measured. The tip can also be used to measure ligand binding by attaching a ligand to the tip. An example how these forces are measured by the tip movement is shown in Figure 1.59.

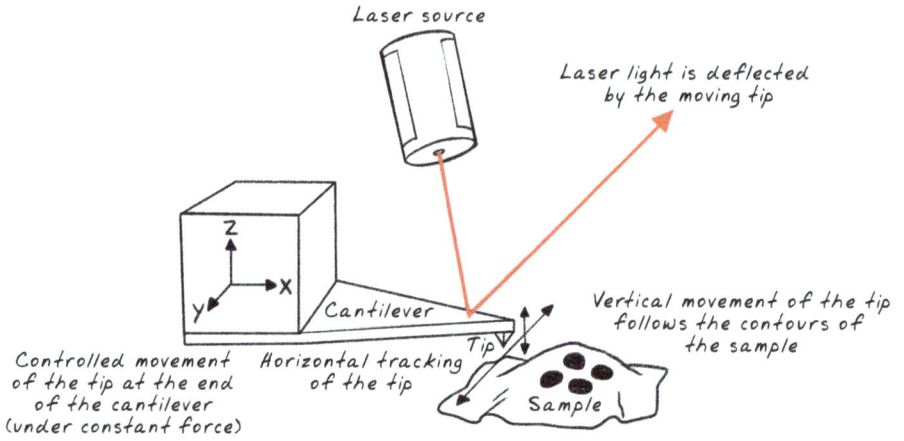

Figure 1.57: The atomic force microscope (AFM) can measure the topography of a surface down to atomic resolution by using the repulsive force between two atoms. That force is measured by the movement of the laser on the back of the tip.

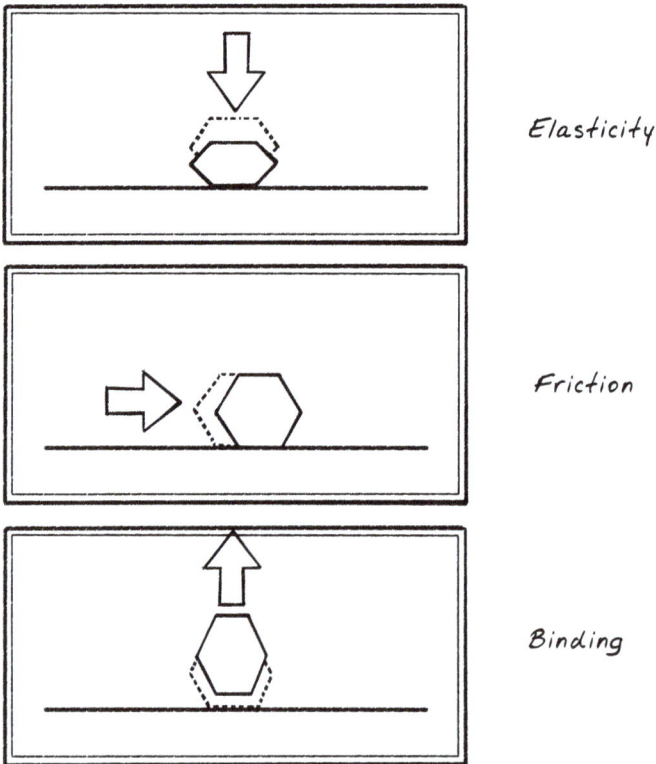

Figure 1.58: These are the different forces an AFM can measure.

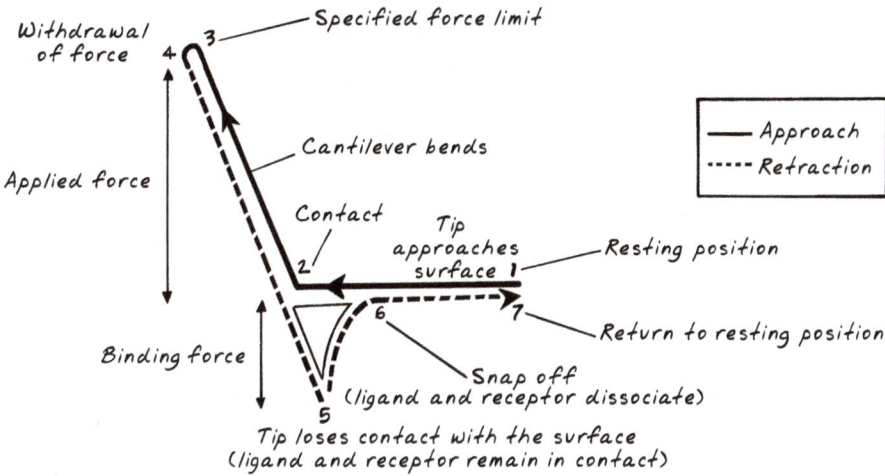

Figure 1.59: The AFM can measure forces on the nanoscale by comparing the adhesion and retraction forces of the tip while keeping track of distances in the z-direction accurately.

There is another method I need to mention when talking about measuring in the nanoscale: microfluidics. When working with solutions in very small diameter tubes, the flow will be laminar, unless the system includes specifically designed intersections that will mix the new solvent/compound in the channel (Figure 1.60). Microfluidics can be used for very precise sequential synthesis, but it is mostly used to detect specific compounds with, e. g., a fluorescent label. With that you can detect, i. e., cancer cells or infectious bacteria, in very low concentrations or identify a specific protein within a cell.

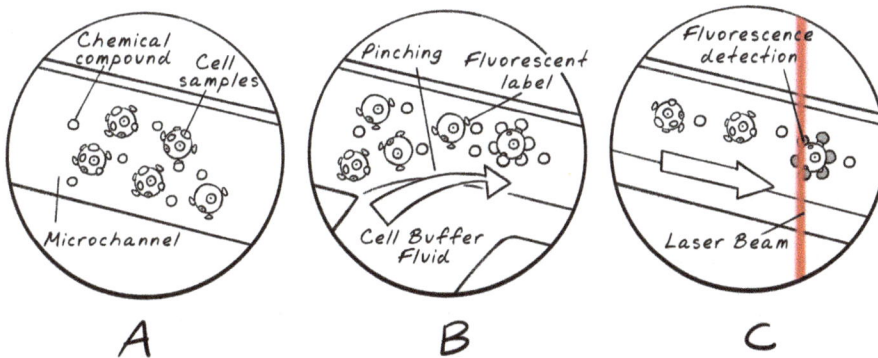

Figure 1.60: Example for a microfluidic testing system. A) Solution with labeled cells is flowing through a very narrow tube with laminar flow. B) The crossing with a junction of a channel containing solvent will dilute the solution and force the cells to flow in a line. C) A laser beam excites the fluorescent label of the cell; the fluorescence will then be measured.

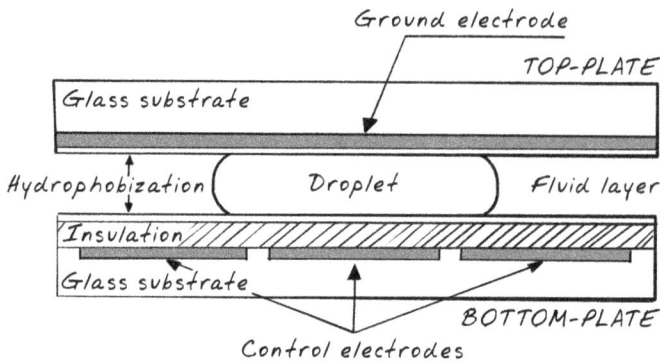

Figure 1.61: Microfluidic system that can manipulate single droplets via electric fields.

A special application of microfluidics uses the small diameter tubes to isolate and manipulate single droplets via an electric field (Figure 1.61). That can be used in synthesis but also in force measurements.

In this book, the analogies between the human and machine processes will be highlighted, not only because they are exciting but also because we can learn from them. The human system has been evolving for millions of years, so the sensor and analysis processes are comparatively simple in structure and function yet powerful and effective in very complex situations. Analogies can highlight how technology can take advantage or mimic these processes, and thus build more effective nanosensors.

The chapters will cover human motion, vision, hearing, smell and taste, and mechanical and heat sensing in skin. In each chapter, we will first look at how human motion or the human sense works on the molecular scale. Then, we will explore how technology has used the human molecules of motion and senses in nanotechnology. The last part of the chapter looks at technology that does not use the same molecules, but does mimic the same function through nanotechnology. There is a lot of ongoing research on these topics; here, only examples are presented that highlight different approaches, specifically approaches that remain close to the original structure and function of human senses and motion.

Bibliography

[1] Guyton AC, Hall JE. Textbook of Medical Physiology. 10th ed. Philadelphia, PA: W.B. Saunders Company; 2000.
[2] Vogel S. Cats' Paws and Catapults. New York, NY: W.W. Norten and Company; 1998.

Further Reading

Clark RAF. The Molecular and Cellular Biology of Wound Repair. 2nd ed. New York, NY: Kluwer Academic Publishers; 1996.

Garg HG, Longaker MT. Scarless Wound Healing. New York, NY: Marcel Dekker, Inc.; 2000.

Grattarola M, Massobrio G. Bioelectronics Handbook: MOSFET's, Biosensors, and Neurons. New York, NY: McGraw-Hill; 1998.

Kandel ER, Schwartz JH, Jessell TM. Principles of Neural Science. 4th ed. New York, NY: McGraw-Hill; 2000.

Laurencin CT, Nair LS. Nanotechnology and Tissue Engineering. Boca Raton, FL: CRC Press; 2008.

Martin DK. Nanobiotechnology of Biomimetic Membranes. Springer; 2007.

Meyers MY, Chen PY. Biological Materials Science. Cambridge University Press; 2014.

Painter PC, Coleman MM. Fundamentals of Polymer Chemistry. 2nd ed. Boca Raton: CRC Press; 1997.

2 Movement

2.1 Human Movement and Muscles on the Molecular Scale

In this chapter, voluntary movement is discussed. As with the human senses, human voluntary movement happens in response to a signal. When the signal occurs, the muscle contracts. Depending on the muscle force needed, a different number of muscle fibers contract. In most cases, this contraction takes place over a short time period and the muscle will quickly return to its extended, relaxed state.

Molecularly-speaking, how does a muscle contract? A muscle contains bundles of parallel muscle fibers called fascicles (Figure 2.1a) [1]. Each muscle fiber is actually a cell with its own cell membrane and nucleus, as well as storage granules containing glycogen (see 1.2, *Structure and function of molecules – sugars and polysaccharides*). The muscle cell has several important special features: it contains the sarcolemma, which is a large, membrane-covered storage space for calcium ions (in fact, there are basically no calcium ions present in the cell with the exception of the calcium ions in the sarcolemma). Additionally, the cell membrane of muscle cells is special in that it is charged, and it can change its charge as a nerve cell does by pumping sodium and potassium ions in and out of the cell.

The signal for contraction comes from nerve cells originating in the spinal cord. If these nerve cells release their neurotransmitter acetylcholine, which activates ion channels in the cell membrane, ions will be released and thus the charge on the membrane is changed (depolarization) (Figure 2.1b) [1]. As soon as the charge of the membrane changes, calcium channels in the membrane are activated to pump a small amount of calcium ions into the muscle cell. Those few calcium ions are sufficient to activate calcium channels in the sarcolemma, which then pump many calcium ions into the muscle cell. As seen with many signals in cells, the stepwise activation leads to the amplification of the signal and thus a fast change, as needed for voluntary muscle contraction [2]. Now that we have the signal, how does the signal lead to an actual force?

To understand that, another special set of features of the muscle cells must be explained: the myofilaments in the myofibrils, actin and myosin (Figure 2.1). Actin is a fiber that is stiff and fixed in the cell. Myosin is also a stiff fiber but has a lot of heads that can move [3]. Myosin can walk with those heads along the actin fibers and pull the whole fibers and thus cells with it, 10 nanometers at a time (Figure 2.2). This occurs via lever action, part of the myosin head protein being built like a lever [4–6]. So each muscle contraction is a combination of a lot of concurrent 10 nm lever actions all parallel to each other and in the same direction [7].

When the muscle is at rest, the actin fiber is covered and does not allow the myosin to bind [1]. Calcium ions in the muscle cell essentially pull the covers of the actin fibers away, exposing binding sites for the myosin heads. The myosin heads are always activated, i. e. ready for the next pull, when the muscle is at rest. Therefore, as soon as it is possible, the myosin heads will bind the actin fibers and move the head so that the

https://doi.org/10.1515/9783110779196-002

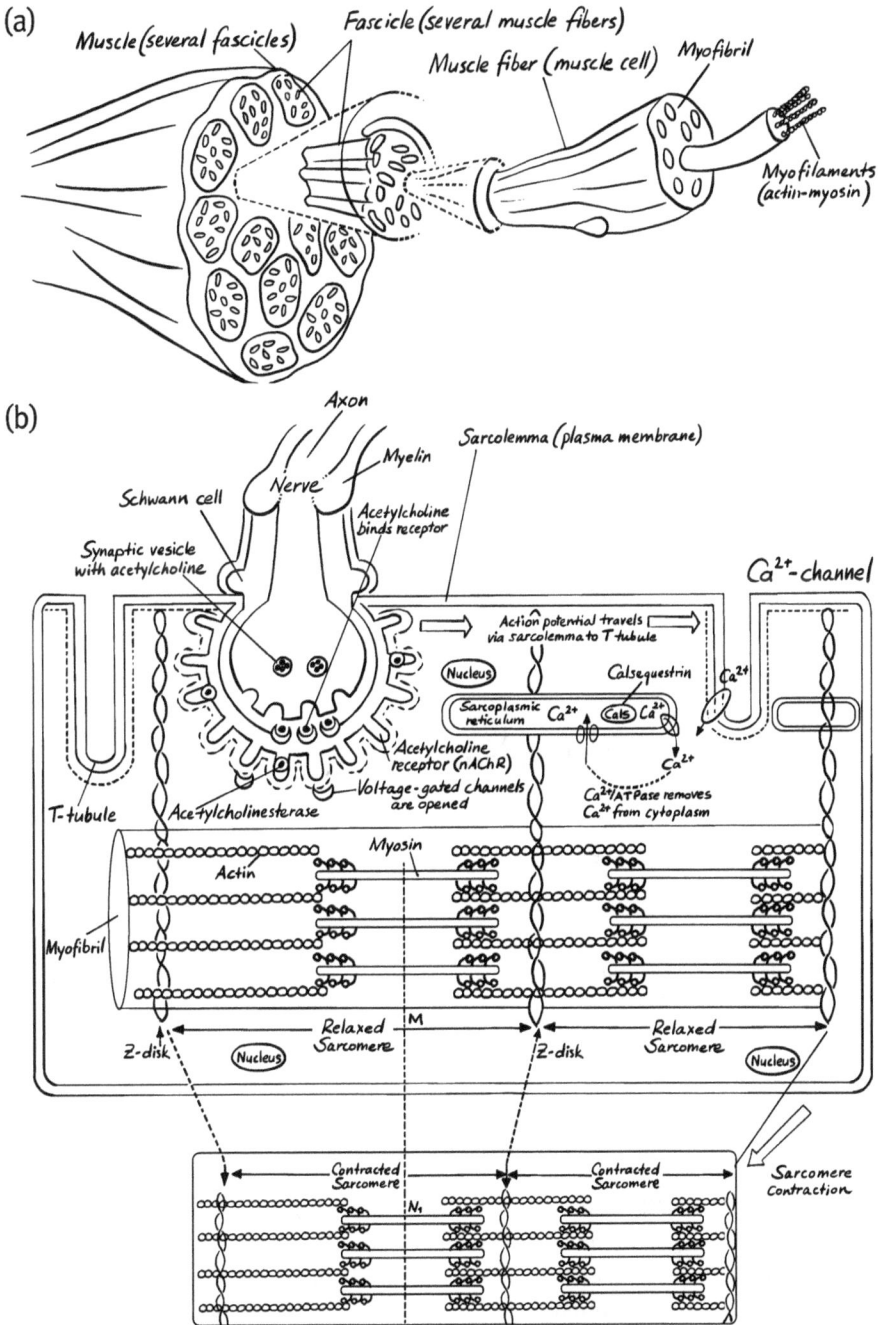

Figure 2.1: (a) Muscle structure and (b) muscle contraction from the human to the nanoscale (adapted from [1]).

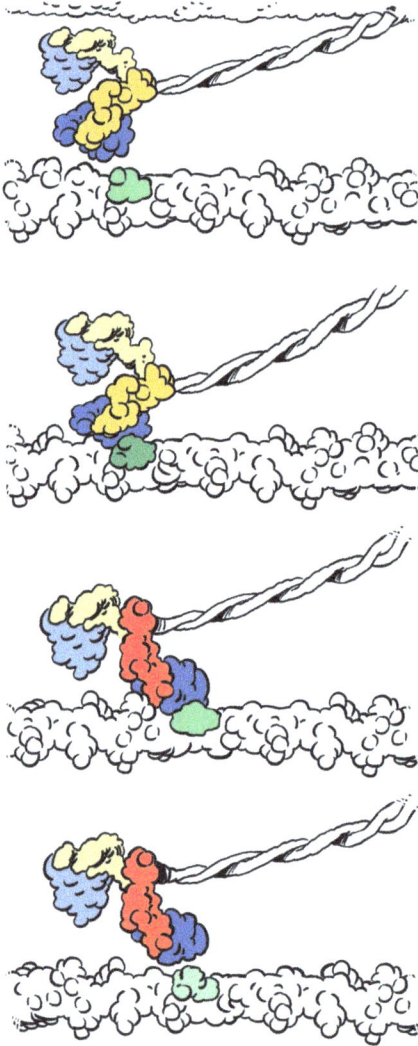

Figure 2.2: The myosin head walks along the actin fiber, acting as a lever and thus creating the force in muscle contraction (adapted from [5]).

muscle fibers get shorter. After that, the energy of the myosin head is spent. It will be immediately recharged while the muscle is at rest.

What kind of energy are we talking about? In the body, energy comes from oxidizing the food molecules, sugar (glucose) and fat (Figure 1.31). Usually, oxidizing chemicals results in heat (a fire is a rapid oxidation that can be prevented by blocking access to oxygen). In the body oxidation results in specific controlled chemical reactions that generate reducing agents (nicotinamide adenine dinucleotide, NADH) (Figure 1.30) and adenosine triphosphate (ATP) (Figure 1.29) for future reactions. So essentially, NADH and ATP are the short-term energy storage molecules in the body. NADH can be used in reductions, which generate the different chemical compounds needed for building cell parts. The phosphate from ATP can be transferred to a variety of compounds in the cell, generating a bond so high in energy that with its release another high-energy bond, and thus another needed compound, can be made. During the muscle contraction, a phosphate is added to the myosin head, and its release allows the head to move 10 nm, creating the force [8]. To be recharged, another phosphate is added to the myosin head.

Muscles require a lot of energy, which is where the glucagon granules come in. Glucagon is easily split into its glucose components, and glucose is oxidized through just a few reactions, generating ATP quickly (Figure 1.31). Fat is a lot slower; it is stored in special fat cells far away from the muscle cells; signals thus have to travel via the bloodstream to the fat cells first, and then fat or its components must travel back via the blood stream, where any cell on its way can take some of that energy. This process would be too slow and too diffuse to work for the quick-acting muscle cells. Therefore, the two different energy storage molecules actually have different functions in the body.

Any signal that needs to act fast must also be short in duration. The cell accomplishes this by destroying the acetylcholine very quickly, as well as by pumping the calcium back into the sarcolemma immediately. This removes the signal on both ends of the amplification chain simultaneously.

In summary, muscles contract because many high-energy myosin heads pull forward 10 nm with force along a stiff fiber, actin. The energy of the myosin head comes from the short-term energy-storage molecule ATP and is replenished as soon as it is used up. This occurs in response to signals that start fast and end fast.

There is another type of motor movement on the molecular level: the transport of cargo (organelles, chemical compounds) within cells (Figure 2.3) [9]. The outcome of these two motor systems is very different: carrying an organelle from one end of the cell to the other takes place on the nanoscale, while a muscle contraction and force generation takes place on the meter scale. Nevertheless, the actual molecules and force generation mechanisms are very similar [9].

For cellular transport, a motor protein "walks" on a microtubule, carrying its specific cargo. The walk of the motor consists of one of the stiff protein subunits binding to the microtubule and using force to move for a short distance, using up ATP energy in the process [10–13] (Figure 2.4, A-E) [10]. When the energy is used up, the subunit releases the microtubule via a hinge, moving to the next location on the microtubule ("step"). The process then repeats. Both kinesin and dynein have two subunits involved in this stepping, so that one is always connected to the microtubule. Those steps, though, might be coordinated in a different way [10]. (Figure 2.4, F-G).

The cargo might have to go into a different reaction, and this is possible: the microtubule is actually continuously made or polymerized on one end (the "+" end) and broken down or depolymerized on the other end (the "–" end). There are specific motors that will only go from the + to the – end and others that work the other way around

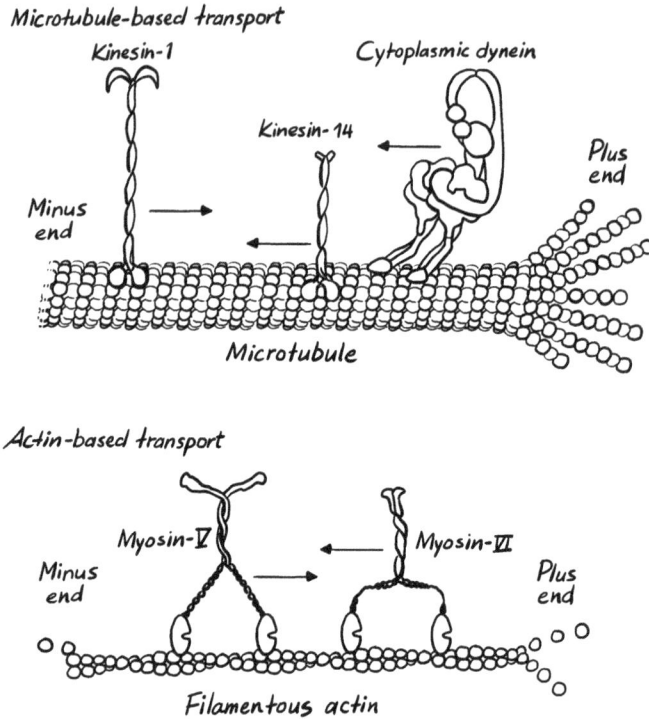

Figure 2.3: Motor proteins also transport cargo (organelles, chemical compounds) in the cell (adapted from [9]).

[9] (Figure 2.3). There are different motor proteins for different cargo. Kinesins transport intracellular organelles, move parts of the spindle during spindle formation, and separate the DNA strands in mitosis and meiosis [14]. Dynein, on the other hand, is responsible for sliding microtubules against one another to generate ciliar and flagellar movement (axonemal dyneins), or for most minus end-directed cargo transport along microtubules (cytoplasmic dyneins).

In summary, transport occurs on microtubule tracks with motor proteins carrying cargo and moving it in either direction. The kinesin and dynein motor proteins work in a very similar manner as actin does during muscle contraction: the "stroke" that moves the fiber forwards is performed by a stiff protein subunit that is powered by ATP. The subunit then detaches and moves to the next site of the fiber, thus taking a "step" of 8–10 nm. This process proceeds across the fiber.

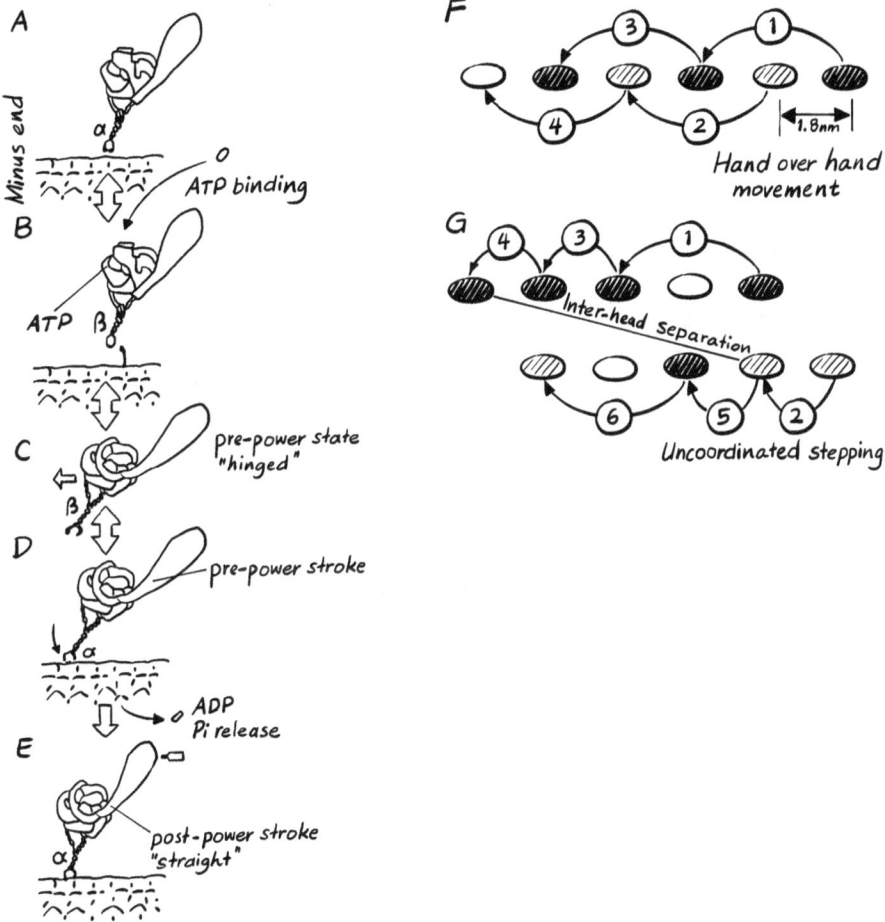

Figure 2.4: The mechanism of cellular transport on a microtubule. A-E: a stiff protein subunit binds to the microtubule, moves for a short distance using force, using ATP as energy. The subunit then releases, moving to the next location, completing a step. F-G: Different motors coordinate the steps in different ways (adapted from [10]).

2.2 Movement Using Biological Molecules and Methods

Figure 2.3 [9] suggests the most common use of biological motor proteins: carrying cargo in a controlled manner on a track in the nanoscale. The track is the actin or microtubule fiber, the motor protein (myosin, kinesin, and dynein are the most common ones, but bacterial motors are used as well) should be controllable, and the cargo can be anything with the right size scale that binds to the motor proteins [9] (Figure 2.5). Ideally, the system should be easy to assemble and then remain stable.

This is where things become a bit more complicated: The track needs to first be assembled somehow. The cell does this actively when needed in the directions needed;

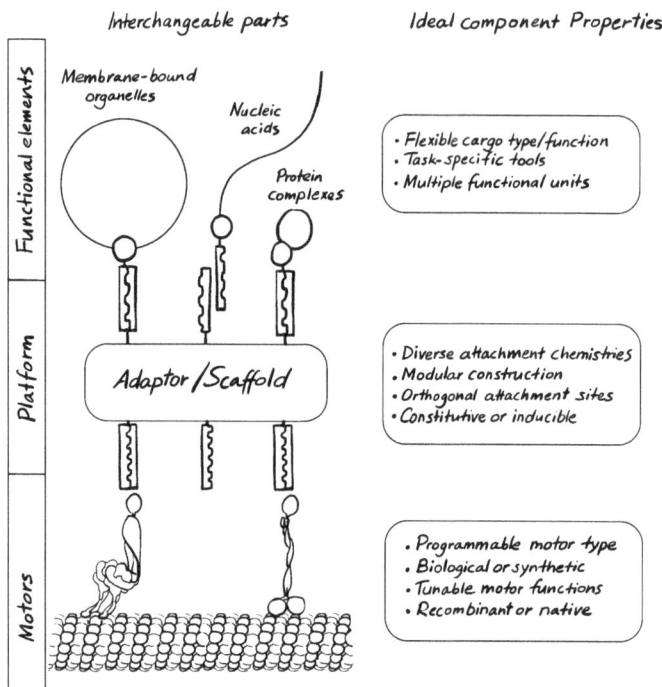

Figure 2.5: Design of nanoscale transport systems based on biological motor proteins (adapted from [9]).

these needs are communicated by chemical and mechanical signals [1]. In artificial systems, this process needs to be controlled, and ideally, in the future, automated. How can we control fiber synthesis, composition, and stability?

Different groups have found different solutions to this problem. Actin monomers self-assemble into linear fibers, and depending on the exact protein composition and salt composition in the solution the fibers might be of different length or might even aggregate into fiber bundles [15]. These fibers or bundles can be aligned by shear forces. For further stability, these fibers can be chemically cross-linked.

Tubulin can be prepared in the same way to create microtubules [15]. In this case, however, the fibers are continually polymerized on one end and depolymerized on the other while transport happens, so the system has to be able to handle (or welcome) this.

A very accurate but difficult way to create the "streets" is using "tweezers" made from a MEMS (micro-electrical-mechanical-systems) chip: due to polarity differences in the microtubule as well as the chip the "tweezers" pick up a microtubule and drop it at a precise location [16].

The transport system can also be inverted: the motor proteins can be attached to a surface and then act as the track on which actin fibers or microtubules move. Motor protein attachment often occurs via micropatterned (or nanopatterned) surfaces with hydrophilic and hydrophobic regions [17]. Usually, the proteins are attached to the bottom of a linear groove that guides microtubule movement. In such a set up, microtubules can even "climb up" walls up to 286 nm.

If there is no groove, a different guiding system is needed. One example is the attachment of carbon nanotubes that are modified with motor proteins, and thus the track for the microtubules [18]. The carbon nanotubes can be aligned by applying a voltage. A nanopatterned surface combined with this specific track allows the microtubules to move past corners and curves. The speed of the movement is based on the ATP concentration, but the direction is random.

One example used kinesin as the ATP-powered motor and microtubules as the carrier (Figure 2.6) [19]. The microtubules carried multilamellar vesicles. Due to the movement of the microtubules several vesicles hit each other with enough force to fuse into a lipid tube network several hundred micrometers in size. These tubular structures mimic the cellular endoplasmic reticulum (ER) and can be used to isolate and trap nanoparticles, as the ER does [19].

How can the cargo be attached? Genetically engineering the motor proteins to add a functional group that reacts easily is one possibility [9]. Equally, a microtubule could be modified with a functional group to attach cargo to it [20]. Biotin-streptavidin, two proteins that bind to each other strongly, have also been used for attachments in various systems.

Now, some sort of traffic control has to be created and the synthesis of the system must be automated. This is obviously the hardest part. A combined microtubule-actin system took the first steps toward traffic control [21]. Since there are two types of tracks, several motor protein shuttles can walk at the same time. The traffic control comes from using motor proteins that walk in opposite directions. It does not seem to be a major problem for them to move past each other, as they do in life cells. That mechanism, though, is not well understood and has limits.

The first reported automatic assembly of a motor protein transport system used a lab-on-chip [20]. The different components of the system were added to different wells, and the microfluidic system operated them in order to prepare the track, attach the cargo onto the motor, and then move the motor onto the track (Figure 2.7). This system allowed them to sequentially attach two different cargos as well.

Moving cargo around is not the only possible application for molecular motors. An intriguing example for other applications is the assembly of a nanosized force meter [22] (Figure 2.8). Here, a microtubule is attached to the end of a bead and fixed there. Another microtubule walks around on kinesin molecules. When the two meet, the fixed microtubule is bent and the resulting force can be measured (it is in the range of piconewtons).

All of these applications are still on the molecular or nanoscale. The body uses its nanosized motors and builds across scales to produce muscle tissue. A similar self-organized system was used to build artificial cilia (Figure 2.9) [23]. Fluorescence-labeled microtubules were attached to polystyrene beads. Kinesin motors were attached to microtubules. When ATP was present, the kinesin motors walked along the microtubules,

(a)

(b)

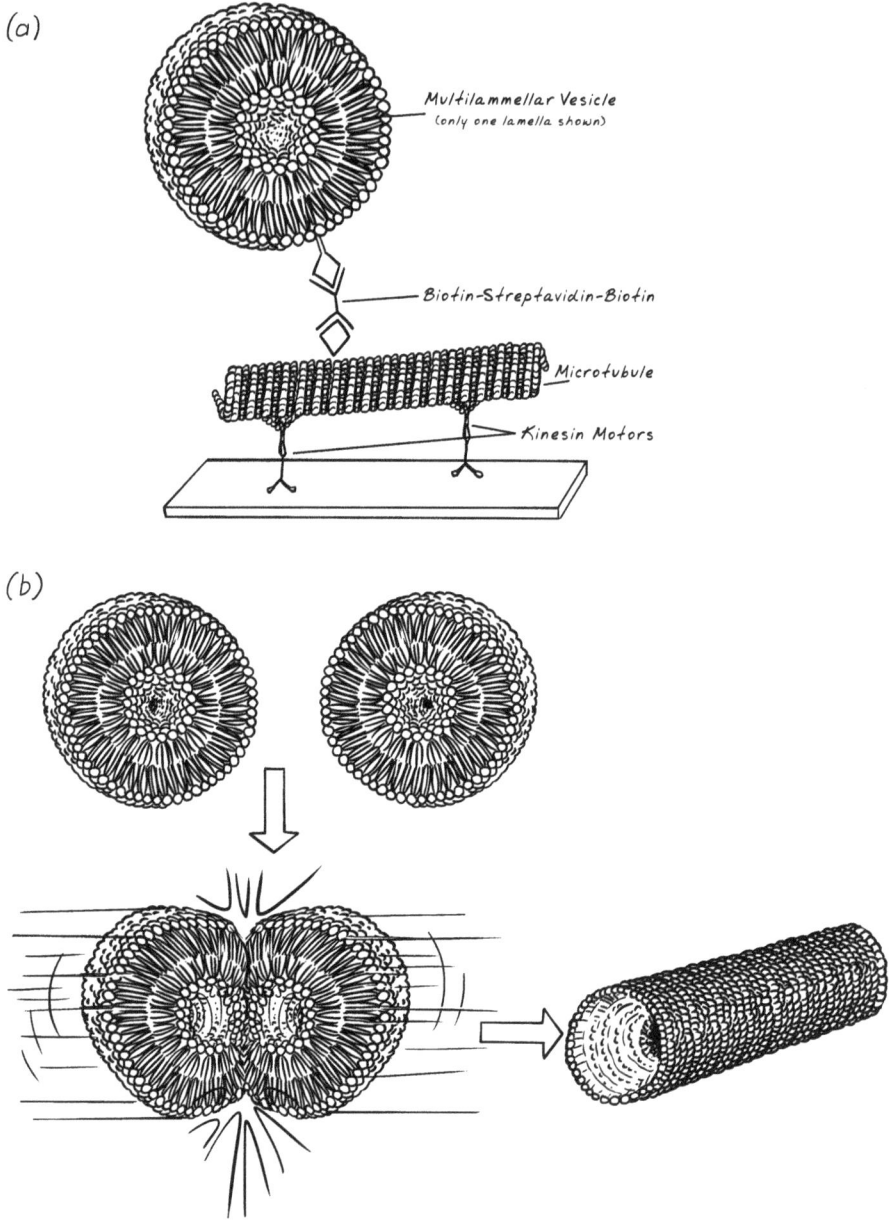

Figure 2.6: (a) Kinesin motors on a surface create a street and move attached microtubules. The microtubules carry multilamellar vesicles. (b) The kinesin can move the microtubules, and with that the vesicles move so fast that the vesicles hit each other with enough force to fuse into a lipid-tube system that mimics the cellular endoplasmic reticulum (ER) (adapted from [19]).

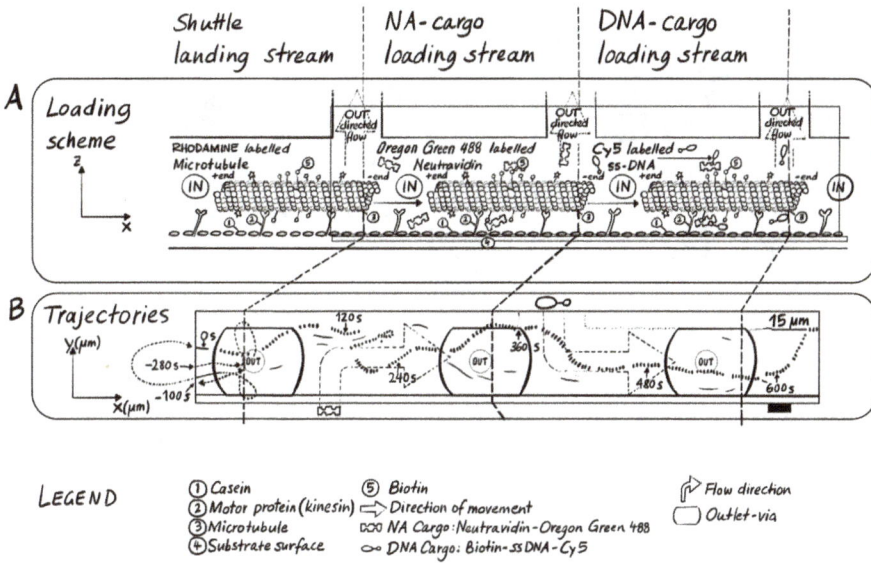

Figure 2.7: First reported automatic assembly of a motor protein transport (adapted from [20]).

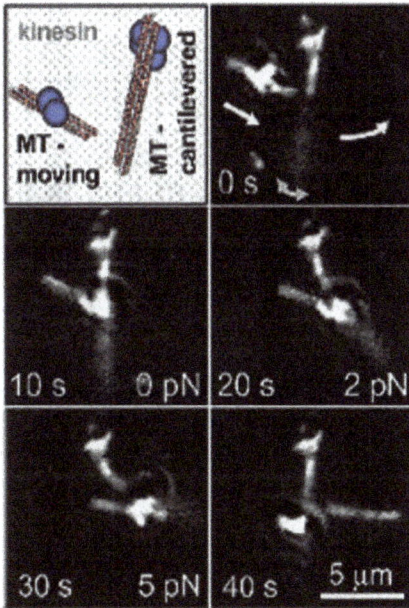

Figure 2.8: Nanosized force meter based on motor protein [22].

which produced a flagella-like motion that could be monitored via fluorescence. Methyl cellulose was used to control the viscosity of the solution and to measure the beating strength of the artificial flagella.

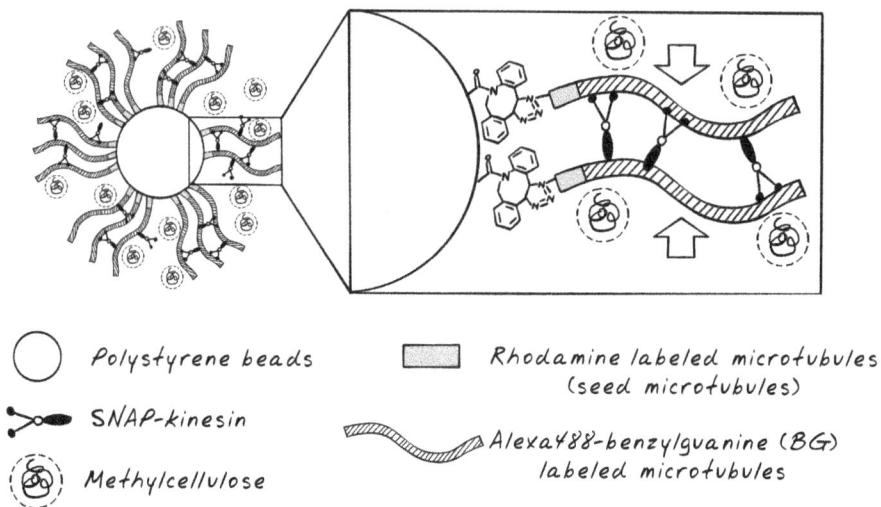

Figure 2.9: Microtubules are attached to a polystyrene bead. Kinesin is attached to the fluorescence-labeled microtubules via SNAP proteins. When the kinesin motors walk along neighboring microtubules, a flagella motion develops that can be monitored by fluorescence (adapted from [23]).

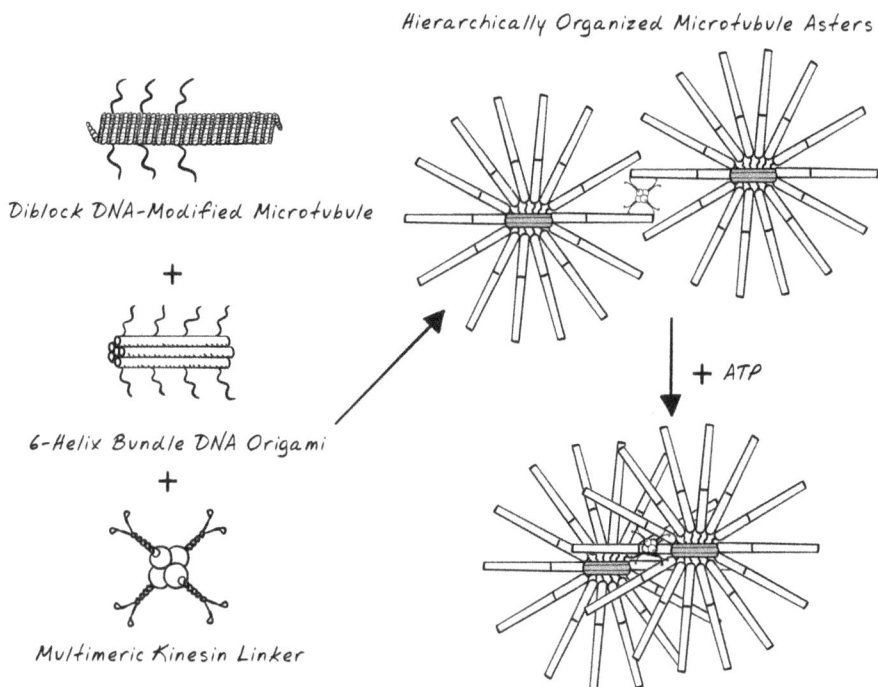

Figure 2.10: Microtubule "asters" are held together by DNA bundles. Multimeric kinesin linkers then walk on the microtubules, thus contracting the artificial muscles (adapted from [24]).

A similar approach can be used to scale up artificial muscles (Figure 2.10) [24]. This is another kinesin-microtubule microtubule/kinesin system that self-organizes from the nanosize to tens of millimeters. Microtubule structures called "asters" (after the flower) are held together by DNA bundles. These self-assemble via single-stranded DNA into larger-scale structures. Multimeric kinesin linkers then walk on the microtubules in the presence of ATP, thus contracting the "artificial muscles".

To summarize, most applications of molecular motor proteins take advantage of their movement. Several methods have been developed to control the direction of their movement, even when more than one cargo is moving at the same time. An automatic, more general assembly has now been reported as well. Cargo attachment, however, is still a rather involved process, and large-scale preparation or long distance and long term transport are still difficult. Initial work has been done to create self-organizing structures across scales to create microscopic artificial flagella and muscles.

2.3 Biomimetic Movement

There are currently different approaches that use molecules for movement. One interesting approach is to use random Brownian motion as the driving force (e. g., [25, 26]). Another approach uses catalysts (e. g., [27]). These methods have not achieved a purposeful direction or transport of other molecules yet and, therefore, will not be described in detail here.

Instead, we will select approaches that are based on small molecules and have already achieved some movement control. The first approach works with rotaxanes. Rotaxanes are molecules that consist of two parts: a dumbbell-shaped molecule and a free-floating ring around the dumbbell that is small enough in diameter that it cannot escape at the ends of the dumbbell (Figure 2.11). Movement occurs when the ring moves from one end of the dumbbell to the other. This can be controlled via several mechanisms. In one example, this movement is initiated by a redox reaction on part of the dumbbell [28] (Figure 2.11). The ring consists of positively-charged, aromatic compounds. The dumbbell contains an electron-rich and a neutral aromatic site. The ring will reside on the electron-rich site and will only move after that site has turned positive via a redox reaction. This is a reversible process, thus the redox reaction acts as a switch. This movement could be used, e. g., for transport if the ring can be fixed to a surface.

The group of Huang has also worked with a rotaxane containing two rings [29] (Figure 2.12).

The two rings of the rotaxane are connected to an AFM cantilever. With an oxidation, the two rings move from the outside to the inside of the molecule, thus bending the cantilever (Figure 2.13). It has been demonstrated that the movement of the rings can be translated into a force, in this case bending a cantilever. Unfortunately, after 20 cycles or so the rotaxane degrades and the bending stops.

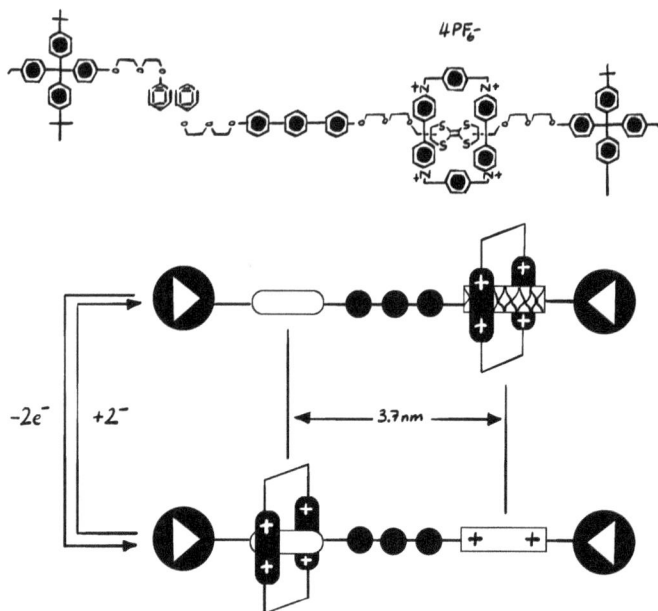

Figure 2.11: Controlled movement of a ring of a rotaxane in a specific direction (adapted from [28]).

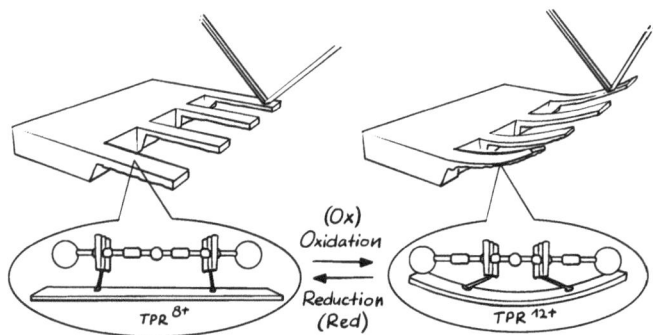

Figure 2.12: A rotaxane containing two rings (adapted from [29]).

Another group used a pseudo-rotaxane dimer to achieve linear movement in one molecule [30] (Figure 2.14). Rotaxane dimers including transition metals work by the same principle ([31] and references therein). Other molecules use light energy to move the ring.

Another approach is using self-assembling structures to create movement on the microscale. One example for that is to use interlocking DNA molecules to create tubes in the range of 0.3–3 µm (Figure 2.15) [32]. The piston that is moving within the tube can be monitored using fluorescence. Speeds up to 0.3 μm^2/s and up to 3 µm total displacements have been achieved, which is similar to molecular motors.

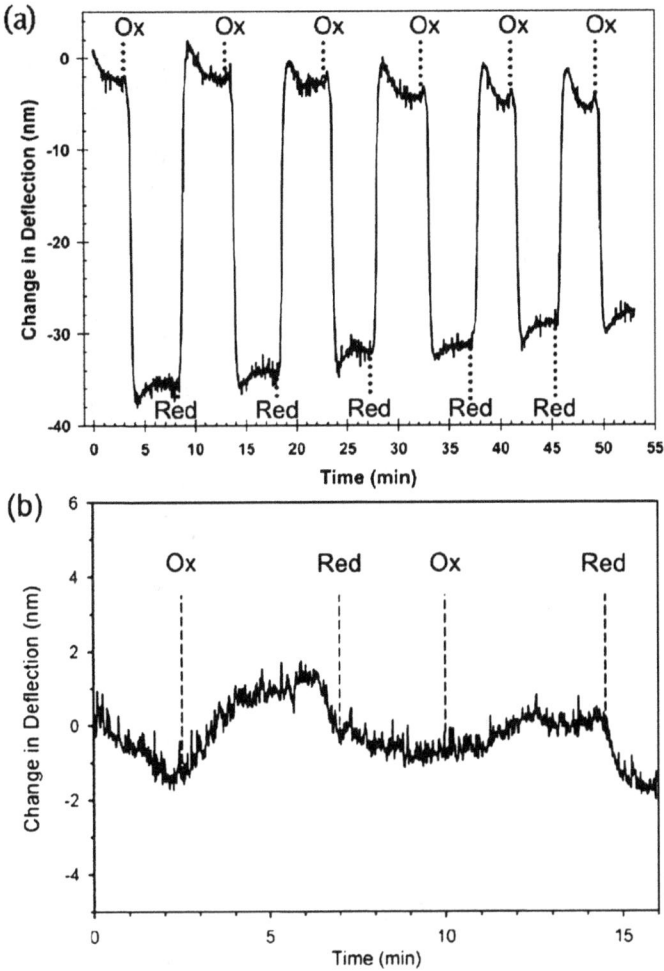

Figure 2.13: The movement of the two rings on the rotaxane in Figure 2.12 is detected by the deflection of the AFM cantilever [29]. (a) Initial cycles, (b) after 20 cycles.

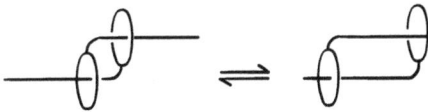

Figure 2.14: Linear movement with a pseudo-rotaxane dimer (adapted from [30]).

In another system, self-assembled asymmetric polymer particles are being propelled via enzymes and a chemical potential (Figure 2.16) [33]. Asymmetric polymer particles ("polymersomes") are self-assembled and incorporate enzymes. In this example, glucose oxidase is combined with catalase. When the substrate, glucose, is introduced, glucose oxidase produces hydrogen peroxide, which in turn is the substrate for catalase, which

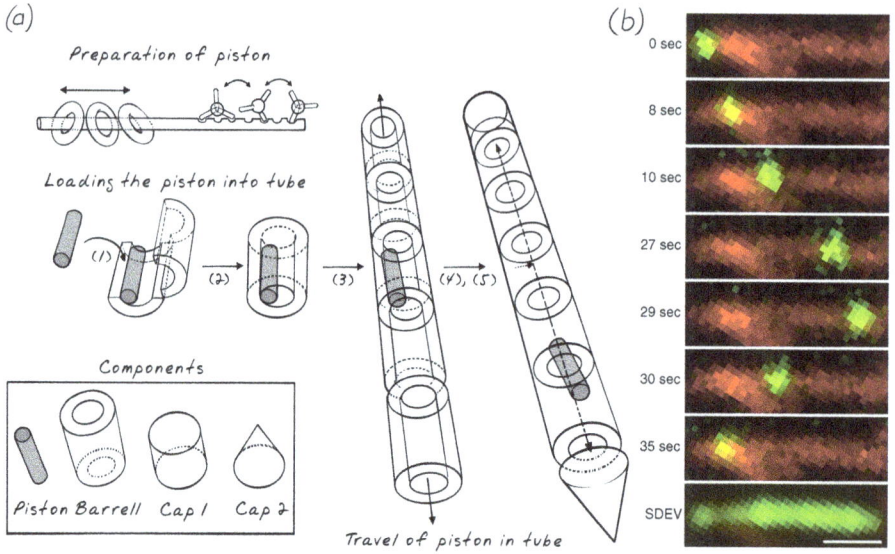

Figure 2.15: (a) Interlocking DNA is used to create tubes (0.3–3 μm) and a "piston" to move within the tubes. (b) Fluorescence is used to show the piston moves within the tube [32].

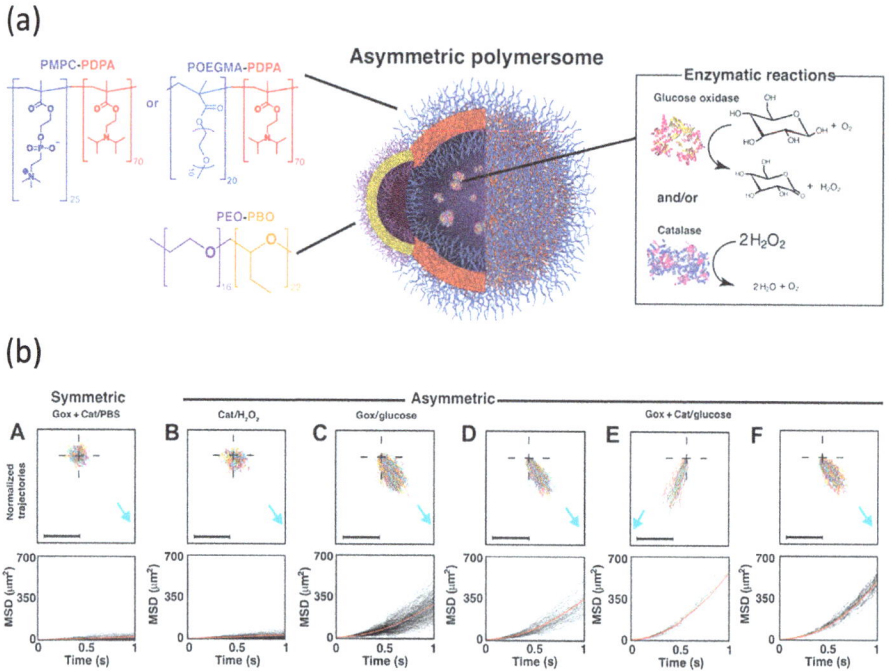

Figure 2.16: (a) Asymmetric polymer particles ("polymersomes") incorporate two different enzymes. When given their substrates, gas will be produced that will propel the polymersome forward. (b) The movement of the polymersome is directional due to its asymmetric structure [33].

will produce oxygen gas. The gas will propel the polymersome forward. The movement of the polymersome is directional due to the asymmetric structure of the particle. This results in movement mimicking chemotaxis and can be used to carry drugs to diseased sites in the body.

2.4 Summary and the Bigger Picture

Human movement on the molecular scale works by the stiff "head" of a molecular motor moving along a fiber as if it were a street. The head movement is fueled by ATP, and controlled by fast concentration changes of ions inside and outside of the muscle cell that contains the walking head. The duration and the strength of muscle contraction can also be controlled in this system. This system self-assembles to create muscles that function on the macroscale.

Motor proteins and their corresponding fiber "streets" have been used extensively in nanotechnology research. It is now possible to automate the assembly of such systems, as well as to control the direction of several cargo "trucks" moving at the same time. It is still difficult to load the motor protein or truck, however. In addition, the problems of large-scale and long-duration movement have not yet been solved. Initial work has demonstrated that some systems can be scaled up to micrometer size by self-assembly.

Biomimetic movement with molecules has not achieved purposeful carrying capacities yet. But an initial self-assembled system allowed for movement in a tube. Another system was able to use a sequence of enzymes to achieve chemotaxis that could be used in drug delivery.

It is even more difficult to mimic motor-movement with other molecules. The only, rather short, "street" that has been shown to give directional, planned movement is the length of a rotaxane-type molecule. Only a few nanometers of movement is possible in these systems. It is also difficult to load cargo onto the ring of a rotaxane.

Bibliography

[1]   Sharma A. In: Medical Biochemistry: Molecules to Disease. iBooks. 2016. https://itunes.apple.com/us/book/medical-biochemistry-molecules/id1119168051?mt=11.
[2]   Krauss G. In: Biochemistry of Signal Transduction and Regulation. Weinheim, Germany: Wiley-VCH; 2014.
[3]   Raymaent I. Trends in Biochemical Sciences. 1994;19:129–134.
[4]   Kull FJ, Endow SA. Journal of Cell Science. 2013;126:9–19.
[5]   Vale RD, Milligan RA. Science. 2000;288:88–95.
[6]   Ikezaki K, Komori T, Sugawa M, Arai Y, Nishikawa S, Iwane AH, Yanagida T. Small. 2012;8:3035–3040.
[7]   Mukherjee S, Warshel A. PNAS. 2013;110:17326–17331.
[8]   Lampinen MJ, Noponen T. Journal of Theoretical Biology. 2005;236:397–421.
[9]   Goodman BS, Derr ND, Reck-Peterson SL. Trends in Cell Biology. 2012;22:644–652.
[10] Kikkawa M. The Journal of Cell Biology. 2013;202:15–23.

[11] Numata N, Kon T, Shima T, Imamula K, Mogami T, Ohkura R, Sutoh K, Sutoh K. Biochemical Society Transactions. 2008;36:131–135.

[12] Seog DH, Lee DH, Lee SK. Journal of Korean Medical Science. 2004;19:1–7.

[13] King SM. Biochimica et Biophysica Acta. 2000;1496:60–75.

[14] Reilein AR, Rogers SL, Tuma C, Gelfand VI. International Review of Cytology. 2000;204:179–238.

[15] Kabir AMR, Kakugo A, Gong JP, Osada Y. Macromolecular Bioscience. 2011;11:1314–1324.

[16] Tarhan MC, Yokokawa R, Jalabert L, Collard D, Fujita H. Small. 2017;13:1701136.

[17] Hess H, Clemmens J, Qin D, Howard J, Vogel V. Nano Letters. 2001;1:235–239.

[18] Sikora A, Ramón-Azcón J, Kim K, Reaves K, Nakazawa H, Umetsu M, Kumagai I, Adschiri T, Shiku H, Matsue T, Hwang W, Teizer W. Nano Letters. 2014;14:876–881.

[19] Bouxsein NF, Carroll-Portillo A, Bachand M, Sasaki DY, Bachand GD. Langmuir. 2013;29:2992–2999.

[20] Steuerwald D, Früh SM, Griss R, Lovchikab RD, Vogel V. Lab on a Chip. 2014;14:3729–3738.

[21] Dong C, Dinu CZ. Current Opinion in Biotechnology. 2013;24:612–619.

[22] Hess H, Howard I, Vogel V. Nano Letters. 2002;2:1113–1115.

[23] Sasaki R, Rashedul Kabir AMd, Inoue D, Anan S, Kimura AP, Konagaya A, Sadaa K, Kakugo A. Nanoscale. 2018;10:6323.

[24] Matsuda K, Kabir AR, Akamatsu N, Saito A, Ishikawa S, Matsuyama T, Ditzer O, Islam S, Ohya Y, Sada K, Konagaya A, Kuzuya A, Kakugo A. Nano Letters. 2019;19:3933–3938.

[25] Lavella GJ, Jadhav AD, Maharbiz MM. Nano Letters. 2012;12:4983–4987.

[26] Hernandez JV, Kay ER, Leigh DA. Science. 2004;306:1532–1537.

[27] Pavlick RA, Dey KK, Sirjoosingh A, Benesi A, Sen A. Nanoscale. 2013;5:1301–1304.

[28] Tseng HR, Vignon SA, Stoddart JF. Angewandte Chemie, International Edition. 2003;42:1491–1495.

[29] Huang TJ. Proceedings of SPIE. 2007;6524:65240H-65241–65240H-65248.

[30] Consuelo Jimenez M, Dietrich-Buchecker C, Sauvage JP. Angewandte Chemie, International Edition. 2000;39:3284–3287.

[31] Niess F, Duplan V, Sauvage JP. Chemistry Letters. 2014;43:964–974.

[32] Stömmer P, Kiefer H, Kopperger E, Honemann MN, Kube M, Simmel FC, Netz RR, Dietz H. Nature Communications. 2021;12:4393.

[33] Joseph A, Contini C, Cecchin D, Nyberg S, Ruiz Perez L, Gaitzsch J, Fullstone G, Tian X, Azizi J, Preston J, Volpe G, Battaglia G. Science Advances. 2017;3(0):e1700362. [Creative Commons Attribution license (CC BY 4.0), https://creativecommons.org/licenses/by/4.0/].

Further Reading

Saper G, Hess H. Synthetic Systems Powered by Molecular Motors. Chemical Reviews. 2020;120:288–309.

3 Vision

3.1 Human Vision on the Molecular Scale

All sensors, biological or technological, have several elements: the sensing element that senses the signal, the transducer that transfers the signal, and an amplification and/or analysis/reporting element that increases the signal and/or analyzes it. In human vision, the sensing element is the eye (Figure 3.1a). On the molecular scale, the sensing elements are specifically the molecules rhodopsin and iodopsin in the rod and the cone, respectively, which are activated by light photons (Figure 3.1b). Rhodopsin and iodopsin contain 11-cis retinal derivatives bound to different proteins called opsins. Light activation turns 11-cis retinal into 11-trans retinal, changing the molecule from a bent, bulky molecule into a long, thin one. The opsin, though, cannot bind the long, thin molecule well anymore, and thus releases it and changes its own conformation in the process. This new opsin conformation fits and binds well to a specific G-protein-coupled receptor called transducin [1, 2] (Figure 3.1b). As the name suggest, this is the initial stage of transducing the signal. In this case, the signal is amplified by a signal transduction pathway that eventually closes an ion channel, which hyperpolarizes the outer cell membrane of the rod or cone. The amount of change in membrane potential is dependent on the amount of light activation and is transferred not via action potentials but as a current in the cytoplasm [3]. This current induces the rod or cone to release less of the inhibitory neurotransmitter glutamate, thus activating the following nerve cell. In the brain, these activated and firing neurons lead to the analysis of the original signal; e. g. with quick scanning and temporal resolution we now understand that we saw a red rose. This analysis might even connect to other neurons in the brain that tell you that the young man giving you the red rose wants to say that he is in love with you.

Let us summarize what happened here: a photoreceptor (11-cis retinal bound to a protein) was activated by photons, which resulted in a change of protein conformation, which started a signal cascade that amplified and transferred the signal to an ion channel, which changed the potential of the cell membrane. This potential change was further transferred to the brain, where the signal was analyzed and recognized as a red rose. Is it possible to use the molecules and methods of the human vision system to make an artificial, molecular-sized photosensor with similar functions?

3.2 Photosensors Using Biological Molecules and Methods

It is not easy to keep native protein structures in an artificial system; in most cases proteins denature, i. e., they lose their specific structure and become random, losing their function in the process. In the specific case of rhodopsin, not only does the native protein structure need to be preserved but the structure must also cycle through two specific conformations repeatedly, which is difficult to achieve.

https://doi.org/10.1515/9783110779196-003

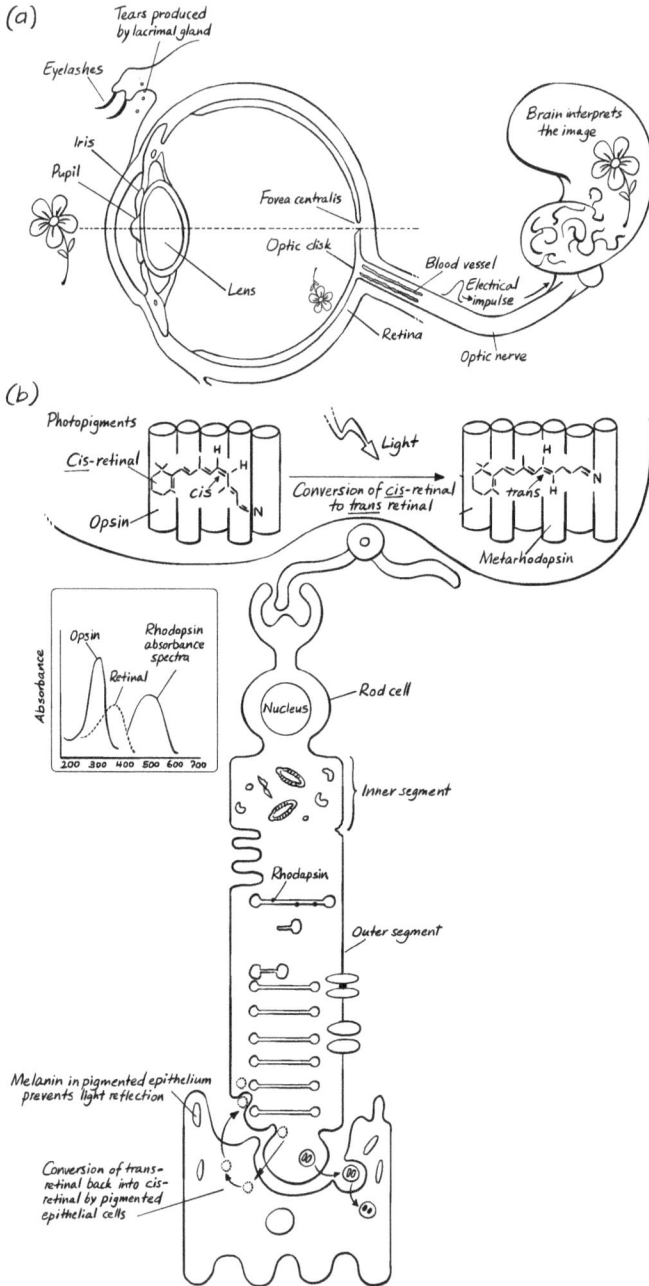

Figure 3.1: The structure of the eye and the molecular process of seeing (adapted from [3]). (a) Large-scale structure of the eye; (b) the photopigment rhodopsin changes its structure when reacting with light, and the change starts a signal cascade in the rod cell.

One group successfully used bacteriorhodopsin, a rhodopsin that can be found in photoactive bacteria called *Halobacterium salinarum*, for a photosensor [4] (Figure 3.2). The sensor uses a photocell with two ITO electrodes, one of them coated with bacteriorhodopsin. An electrolyte solution maintains the native structure of the protein. A laser beam was used to activate the rhodopsin; the laser beam was expanded so that the complete sensing area was irradiated. The amount of light the laser beam emitted was independently monitored via a photodiode. The photocurrent from the bacteriorhodopsin was measured by an oscilloscope and amplified electronically.

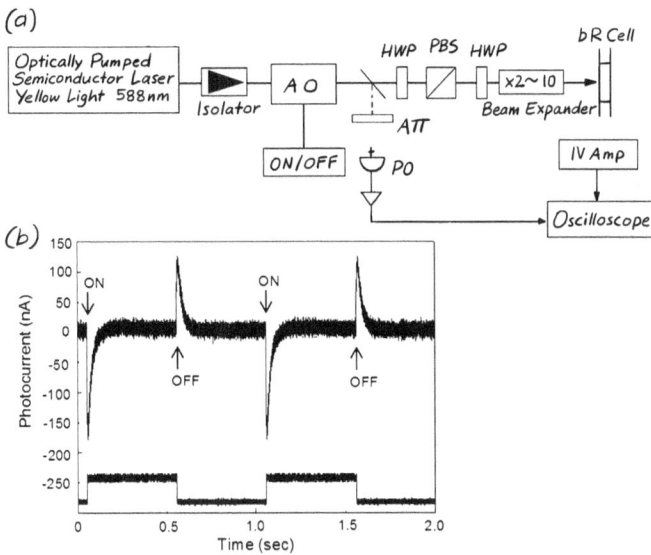

Figure 3.2: Set-up and signal of a photosensor that uses bacteriorhodopsin on an electrode (adapted from [4]). (a) Set-up of the photocell; (b) generated photocurrent based on the amount of light irradiation.

As with the human visual system, this sensor reacts to changes in the amount of light, instead of measuring continuously the amount of light coming from the source. This makes this sensor specifically good at reacting to movement (similar to human vision). When the bacteriorhodopsin film is patterned correctly, the direction of movement can be determined as well. The authors are developing this optical sensor for robots; with this sensor, robots could react to sudden changes of light, and thus dangers [4].

The above sensor used the biological sensing molecule and built it into a common digital photocell. Photocells are used, e. g., in the sensors that turn streetlights on automatically when it is getting dark. But with a biological sensing material functionality can be expanded to detect sudden directional changes and movements, mimicking the function of human eyes instead.

For accurate vision and analysis, the brain moves the eyes to scan an area with a lot of micromovements [5]. These micromovements in scanning were mimicked by another group using a very basic artificial sensor (tiny lens, two digital pixels) [6]. The resulting sensor is more sensitive than related sensors since it detects contrast via movement on top of the number of photons. It was also able to locate a much smaller object than the lens should allow for by looking for contrast. The authors are currently refining this technology so that it can also be used in robotics. It would be powerful to combine both technologies, but nothing to that effect has been reported so far.

Modified rhodopsin can also be used to measure voltage or ions instead of light. To image voltage in life cells in real time, a genetically modified rhodopsin was expressed in neurons or muscle cells in *Caenorhabditis elegans*, a common animal model in neuro and cell biology (Figure 3.3) [7]. It was possible to measure voltage, and with that muscle and neuron activity, *in-vivo*.

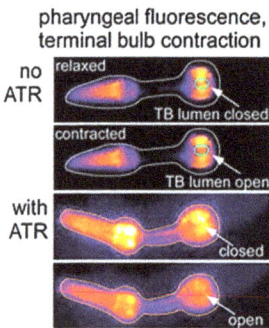

Figure 3.3: All-*trans* retinal (ATR) is used to image intrinsic muscle activity of the pharyngeal muscle under a fluorescence microscope [7].

Iterative directed evolution coupled with protein structure modeling created a rhodopsin variant that measures chloride concentration instead of light (Figure 3.4) [8]. The variant was expressed in E. coli cells that now function as life, in-vivo chloride sensors.

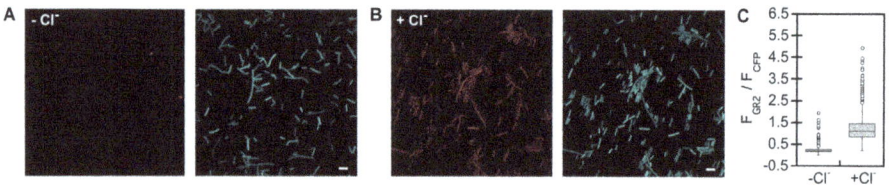

Figure 3.4: Confocal fluorescence microscopy is used to measure chloride *in-vivo* in E. coli cells: (A) 0 mM and (B) 400 mM sodium chloride. For each panel, the emission from a control (red, left) and from the modified resorcinol (cyan, right) are compared (scale bar = 5 μm). (C) Boxplots show the normalized emission response (F_{GR2}/F_{CFP}) of each cell analyzed from four biological replicates (with permission from [8]).

3.3 Biomimetic Photosensors

There are a lot of photosensors that mimic some or all functions of the human vision system (eye combined with brain) with artificial systems: light detection, signal transduction to electric signal, amplification of the signal, and analysis of the signal.

There is a large variety of chirooptical switches, where a specific wavelength induces a change in the three-dimensional structure of the molecule, mimicking the change the retina makes in response to light ([9] and references therein); in this case, the chemical structure changes to a different enantiomer. This change is reversible with a different wavelength light and quite fast. These switches are not connected to any further action or analysis. Instead, they are expected to be implemented as optical computer memory.

Solution processable organic field-effect transistors (OFETs) can also be used as optical sensors called "organic phototransistors" (OPTs) (Figure 3.5) [10]. In this example, the ionic hydrogel containing silver nanowires is a transparent, flexible gate electrode. The pyrrole-thiophene copolymer is a narrow band-gap polymer with high hole mobility that is photosensitive. Therefore, when light hits the copolymer the holes will move toward the drain, generating current, with that transducing a light signal into a current output. Polyimide is an organic semiconductor to complete the transistor. The sensitivity of this photosensor can be tuned, as well as amplified, by the gate voltage.

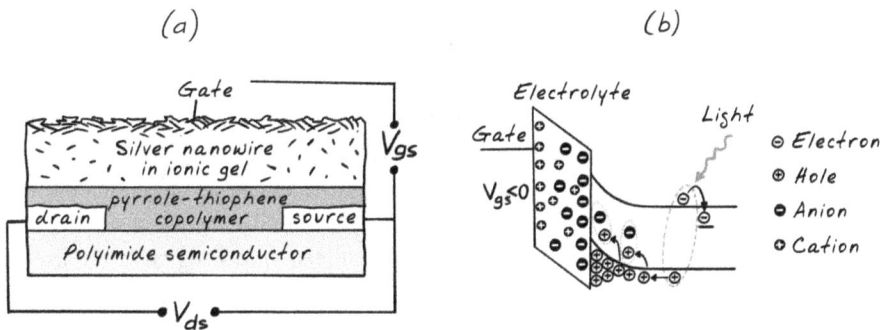

Figure 3.5: (a) Example of an organic field-effect transistor (OFET) that can be used as an optical sensor ("organic phototransistor", OPT) (adapted from [10]). The sensor is the pyrrole-thiophene copolymer, which transduces the light signal into a current output (b). The sensitivity of this photosensor can be tuned (and amplified) by the gate voltage (adapted from [10]).

Another organic phototransistor can react differently to two wavelengths, and thus can be used as an optical logic gate (Figure 3.6) [11]. Dependent on the wavelength, the transistor is an "and" or "or" logic gate, while at the same time still transducing the optical signal to a current output and possibly even amplifying the signal at the same time. These logic gates mimic some of the functions of the brain, where signal output is analyzed by a variety of the following neurons.

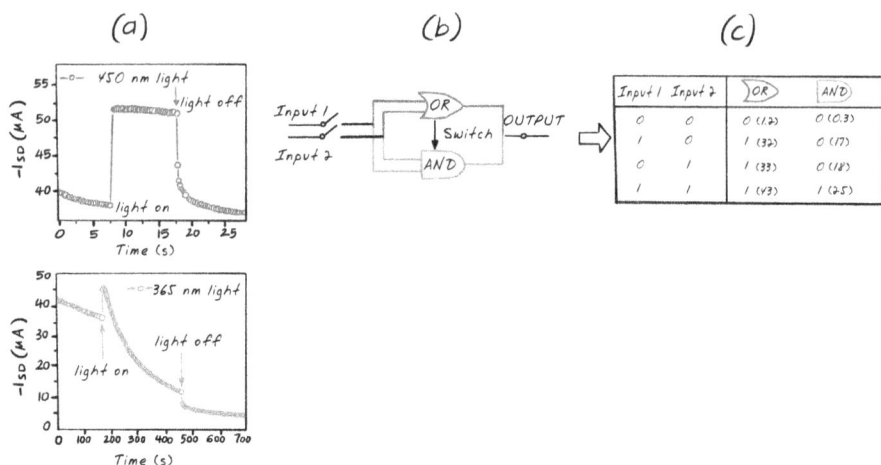

Figure 3.6: (a) Current output for an organic phototransistor that can react to two different wavelengths (365 nm and 450 nm) differently. As with all transistors, this can be used as a switch, but with the different outputs, this one transistor can be used as an "or" or "and" logic gate, dependent on the wavelength that is used (b). (c) shows the truth table for the logic gates (adapted from [11]).

An optical mimic of rods was developed not from a phototransistor but from a modified pyrrole N'-1,N'-6-bis(3-(1-pyrrolyl) propanoyl) hexanedihydrazide (DPH) that self-assembles into rod-like, long structures that are a few nm in diameter [12] (Figure 3.7). These rods are photoluminescent; their signal was enhanced in the presence of a sensitizer. These rods have been effectively used to selectively measure the concentration of pyrrole. So, here the detector is the photoluminescent rod. Detection occurs via changing the optical properties (i. e., quenching the fluorescence) of these rods by changing the excited state energy of the molecule, instead of by changing its three-dimensional structure. The light energy-level is then detected and analyzed.

An example of a chemical structure change coupled with a signal is the use of fluorescence-labeled liposomes as intracellular biosensors [13] (Figure 3.8). Here, a lipid is labeled with the fluorescent dye Nile-Blue and incorporated into liposomes. The liposomes can easily be taken up into cells. A different pH in different cells creates a different protonation state of the dye, which results in a color change (Figure 3.7). Analyzing the specific color results in a measurement of the internal cell pH, and thus acts as an indicator of various diseases (e. g., cancerous cells often have a lower pH). Therefore, this sensor does not measure light, but rather uses light (fluorescence) to measure pH. This is different from the human visual system, where light is detected and the signal is transduced and analyzed as an electrical signal.

These few examples already show that there are endless possibilities to create biomimetic photosensors. Most of them are based on specific compounds binding to a detector surface, thus measuring the amount of a compound in solution, with a light or color output. Only two examples will be highlighted here.

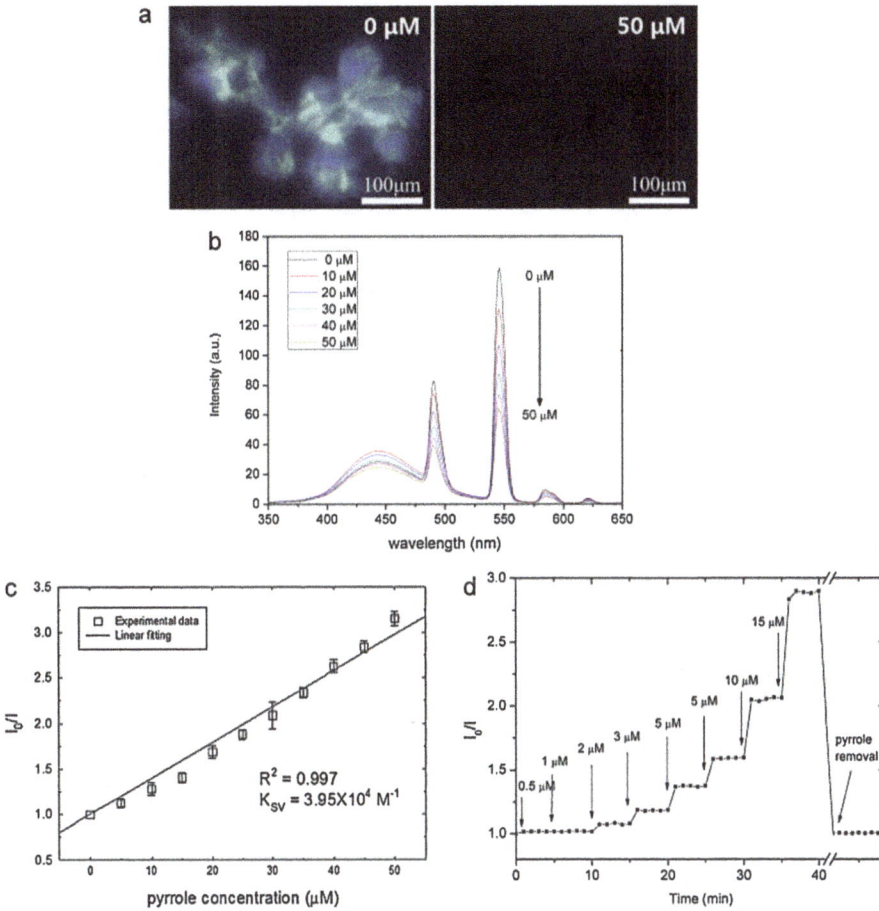

Figure 3.7: A photo-luminescent rod-mimic that detects pyrrole [12]. (a) Luminescence of the rod; (b) different concentrations of pyrrole result in different amount of luminescence; (c) luminescence and pyrrole concentration are linearly related; (d) the luminescence at each concentration is constant over minutes.

In one case, columns are packed with a solid phase of molecularly-imprinted particles (MIPs), whose imprints are specific for antibiotics [14]. A fluorescent analogue to each antibiotic is synthesized. Then a known concentration of a fluorescently-labeled antibiotic will compete with the nonfluorescent antibiotic in binding to the specific, imprinted sites of the column. To determine the amount of antibiotic in a sample, the fluorescence of the solution exciting the column is measured. In this case, the sensing element is then the imprinted site, the transduction is the competing molecule, and the analysis is performed by fluorescent light.

In another example, the binding surface is polyacetylene combined with a lipid monolayer [15]. The color of polyacetylene is based on the conjugation length of

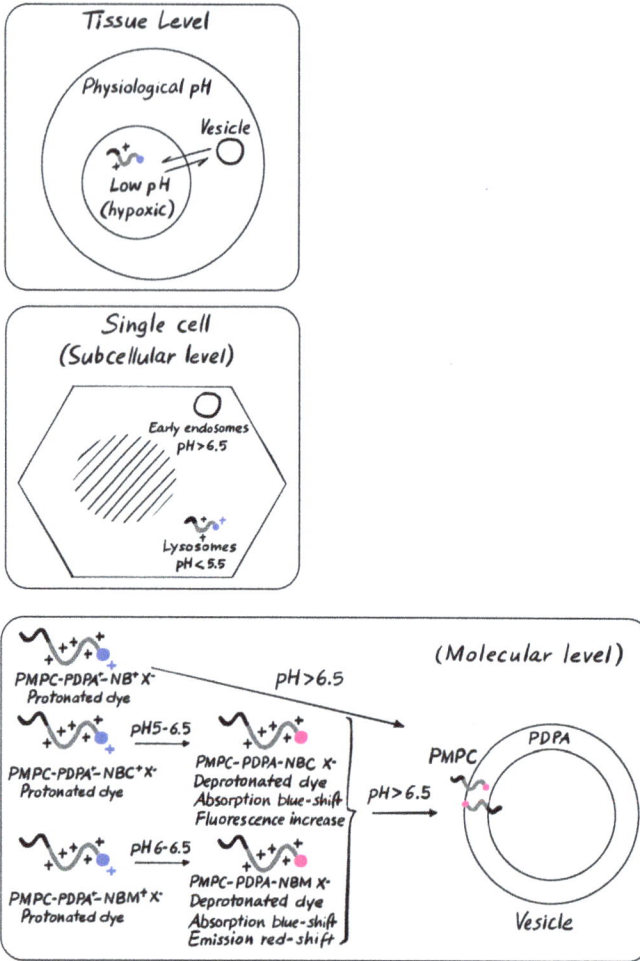

Figure 3.8: Photosensor that uses light (fluorescence) to measure pH (adapted from [13]).

the polymer, which in turn is dependent on how much the perfect alignment of the π-orbitals in the chains is disturbed by bending or stressing the polymer chains through outside forces. In this case, the binding of bacterial products by the lipid monolayer can be detected as a simple color change from blue to red. This technology could be used to develop sensors in the packaging of food that identifies, for example, spoiled meat. In this case, the sensing element and the transducer is the polyacetylene layer, and the signal is a simple visible color change.

3.4 Summary and the Bigger Picture

The eyes are human photoreceptors; detection occurs, however, via a change in the three-dimensional structure of a molecule. This change in the shape of the molecule initiates a change in the shape of a protein, which initiates a signal cascade that amplifies the signal and activates an ion channel, which changes the membrane potential, which initiates the signal transfer to the brain and with that the analysis of the signal. Analysis here does not only imply the recognition of a specific shape such as a rose, but it also means the creation of a connection with the rose to the feeling of love, also using memory in the process.

Using the actual molecules in the process for nanosized photosensors is difficult, since proteins denature and lose function easily in environments other than the natural one. Nevertheless, bacteriorhodopsin has been successfully attached to an electrode, which created a sensor that reacted not to the amount of light itself but the change in the amount of light, as does the human vision system. Rhodopsin can also be genetically modified to measure other things than light, such as voltage or chloride concentration.

Another function of the human system has been mimicked: the constant movement of the eyes to create a more accurate picture and to assist with the analysis of this picture (e. g., by identifying different distances for the various parts of the picture). The combination of both processes should create an even more powerful artificial photosensor.

When developing photosensors on the nanoscale that measure light, the options are endless and a large variety of systems have been reported. They range from chemical or electrical sensors that measure a change in light absorption or fluorescence to the opposite case, where a change in a chemical or current is reported as a color change. What is less common is the connection of sensing to amplification and analysis within one system. Phototransistors have been developed that achieve this. Other examples will be touched on in the following chapter on smell and taste.

Bibliography

[1] Govardovskii VI, Firsov ML. Neuroscience and Behavioral Physiology. 2012;42:180–192.
[2] Kisselev OG, Park JH, Choe HW, Ernst OP. In: Giraldo J, Pin JP, editors. G Protein-Coupled Receptors: From Structure to Function. Cambridge, UK: Royal Society of Chemistry; 2011. p. 54–74.
[3] Sharma A. Medical Biochemistry: Molecules to Disease. iBooks. 2016. https://itunes.apple.com/us/book/medical-biochemistry-molecules/id1119168051?mt=11.
[4] Kasai K, Haruyama Y, Yamada T, Akiba M, Tominari Y, Kaji T, Terui T, Peper F, Tanaka S, Katagiri Y, Kikuchi H, Okada-Shudo Y, Otomo A. Proceedings of SPIE. 2013;8817:88170N-88171–88170N-88178.
[5] Martinez-Conde S, Macknik SL, Hubel DH. Nature Reviews Neuroscience. 2004;5:229–240.
[6] Viollet S, Franceschini S. Sensors and Actuators. A, Physical. 2010;160:60–68.
[7] Hashemia NA, Bergs ACF, Schüler S, Scheiwe RA, Costa WC, Bach M, Liewald JL, Gottschalk A. Proceedings of the National Academy of Sciences. 2019;116:17051–17060. https://doi.org/10.1073/pnas.1902443116 (open access).
[8] Chi H, Zhou Q, Tutol JN, Phelps SM, Lee J, Kapadia P, Morcos F, Dodani SC. ACS Synthetic Biology. 2022;11:1627–1638. https://doi.org/10.1021/acssynbio.2c00033.

[9] Feringa BL, van Delden RA, Koumura M, Geertsema EM. Chemical Reviews. 2000;100:1789–1816.

[10] Xu H, Zhu Q, Lv Y, Deng K, Deng Y, Li Q, Qi S, Chen W, Zhang H. ACS Applied Materials & Interfaces. 2017;9:18134–18141.

[11] Shi Q, Liu D, Dai S, Huang J. Advanced Optical Materials. 2021;9:2100654.

[12] Park S, Lee SY. Sensors and Actuators. B, Chemical. 2014;202:690–698.

[13] Madsen J, Canton I, Warren NJ, Themistou E, Blanazs A, Ustbas B, Tian X, Pearson R, Battaglia G, Lewis AL, Armes SP. Journal of the American Chemical Society. 2013;135:14863–14870.

[14] Urraca JL, Moreno-Bondi MC, Orellana G, Sellergren B, Hall AJ. Analytical Chemistry. 2007;79:4915–4923.

[15] Scindia Y, Silbert L, Volinsky R, Kolusheva S, Jelinek R. Langmuir. 2007;23:4682–4687.

4 Smell and Taste

4.1 Human Smell and Taste on the Molecular Scale

All sensors, biological or technological, contain several elements: the sensing element that senses the signal, the transducer that transfers the signal, and an amplification and/or analysis/reporting element that increases the signal and/or analyzes it. Smell and taste are humans' chemical sensors. The sensing elements for chemicals are G-protein-coupled receptors (GPCR) [1, 2] (see Section 1.1, Figure 1.13). For taste, the taste molecules directly diffuse into the taste pore that contains the GPCR (Figure 4.1) [3–6]. The transfer of the odor molecules to the sensor site is more complicated, since they are a gas and need to be captured first. An odorant-binding protein (ODP) binds the molecule and transports it to the GPCR on the surface of the olfactory bulb (Figure 4.2). Different receptors are concentrated in different areas of the bulb [7]. Sometimes, a surfactant is needed to aid the odor molecule in diffusing to the ODP, so that it can then bind and be transported to the receptor. These surfactants are part of the constant secretion in your nose.

Once the odorant or taste molecule is bound to the GPCR, the receptor is activated and a signal transduction pathway is started; for taste, it is the phospholipase pathway (Figure 1.4), and for smell it is the adenylate cyclase pathway (Figure 1.3). As with all signal transduction pathways, the signal is amplified in the process.

The amplified signals eventually open sodium and calcium channels that depolarize the cell membranes of these specialized nerve cells, thus triggering an action potential. The action potential is processed in the brain via the gustatory afferent nerve in case of taste and the olfactory bulb and olfactory cortex in the case of smell.

Though we have described how the signals travel from molecule to molecule, we have not discussed yet how a few different receptors can lead to the detection and identification of a large number of complex smells and tastes. Some of the details of these processes are not yet known, but principally speaking this happens via each molecule binding to each receptor but with different strengths, and the combination of signals for one molecule results in a more complex sensation [8, 9]. Detection and identification is also combined with memory. Humans avoid chemicals toxic to our body and seek out "good chemicals", i. e., food and clean water, using smell and taste. The pathways in the brain for this memory have been identified [10].

Let us summarize what happened here: odorants and taste molecules bind to G-protein-coupled receptors in a combinatorial manner in the nose or tongue that are part of specialized neurons. When bound, these molecules activate the receptor and start a signal cascade, which then results in an action potential, and is then sent to the processing part of the brain. Is it possible to use the molecules and methods of the human chemical senses and make an artificial, molecular-sized chemical sensor with similar functions?

https://doi.org/10.1515/9783110779196-004

TONGUE

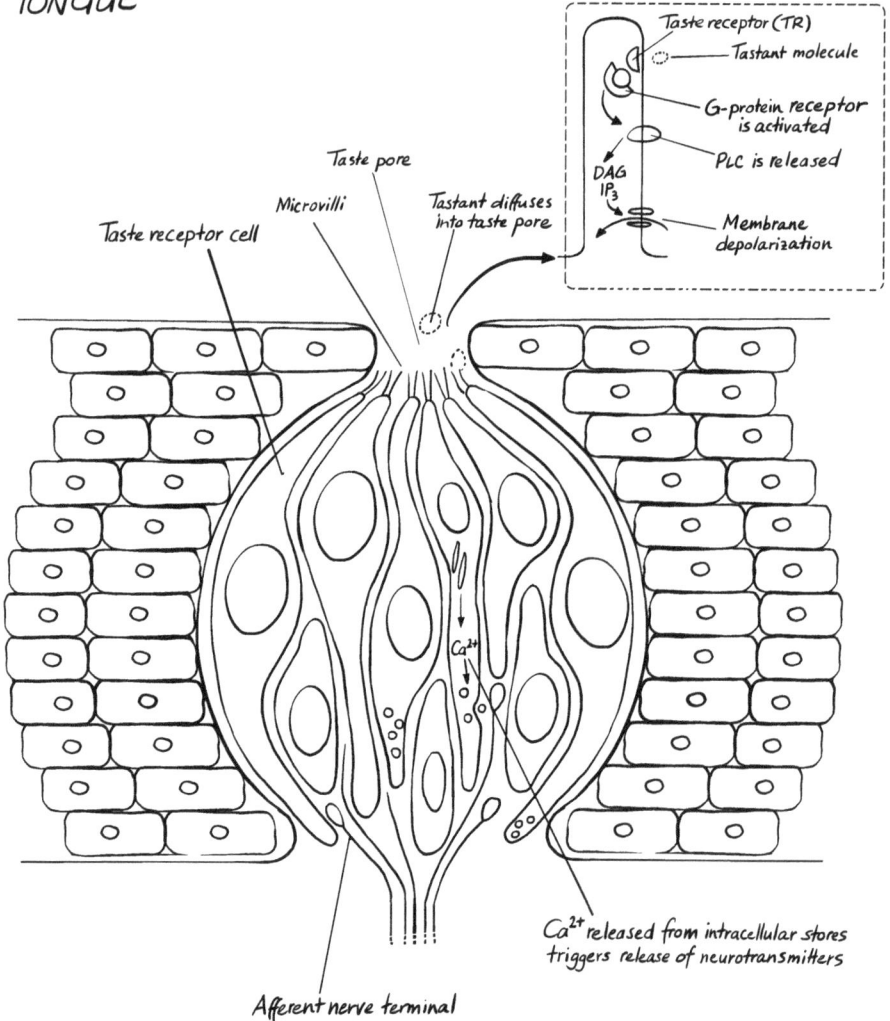

Figure 4.1: The human sense of taste and its molecular actors (adapted from [2]).

4.2 Chemical Sensors Using Biological Cells, Molecules, and Methods

Liu et al. present an excellent overview of the approaches that have been used to build artificial gas sensors out of biological taste and smell cells and molecules—even including sensors with live insects [11] (Figure 4.3).

We will not discuss live insects here, but one interesting approach used insect antennae as the sensor [12]. Different antennae from different insects were used, so that each antennae would create a different signal for one compound, as well as different

SMELL

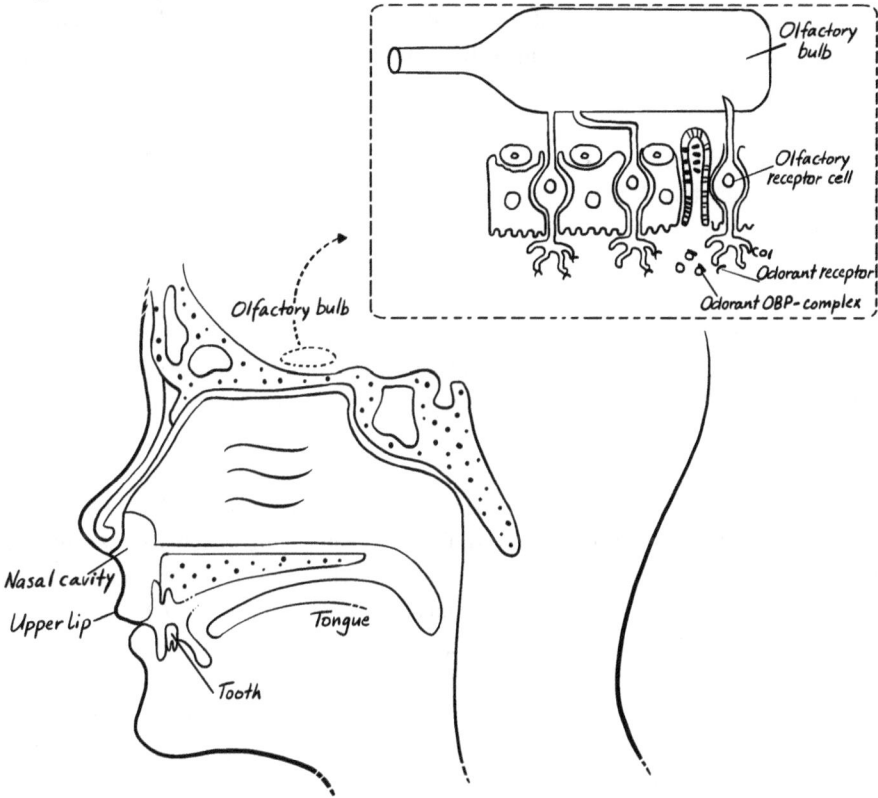

Figure 4.2: The human sense of smell and its molecular actors (adapted from [2]).

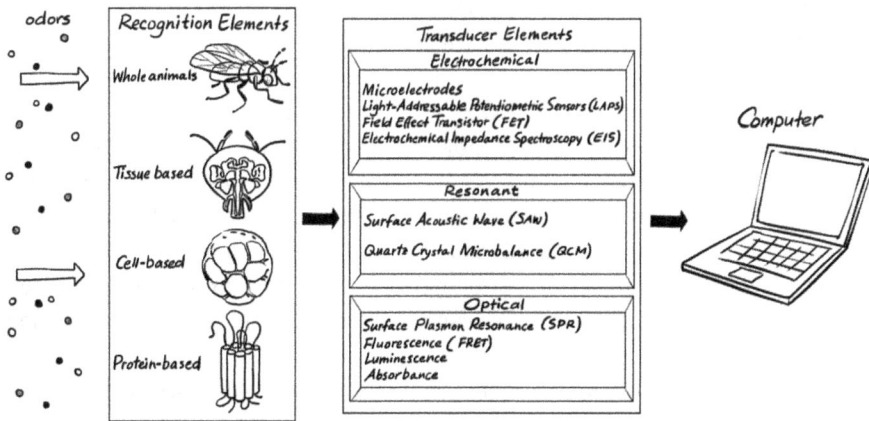

Figure 4.3: Overview about approaches to develop an "artificial nose" (adapted from [11]).

types of signals for different compounds. The signals were detected by measuring an electroantennogram (the electrical activity of an antennae); a program was used to reliably learn the different signals from each antennae and correlate them to the presence of a specific chemical with the four different elecroantennograms for it. After training, the sensor was able to distinguish between eight different odors reliably. Background molecules sometimes interfered with the identification, but not the detection of a compound. Unfortunately, antennae, when removed from the insect, only last 60–90 min. After that, another sensor needs to be built and trained.

A different approach using insect antennae is demonstrated with *Drosophila* flies that can detect cancer by its smell [13]. *Drosophila*'s odor receptors use cAMP as a signaling molecule. These flies are a model organism for genetic engineering; in this case, flies were developed that have cAMP fluorescently marked. The *Drosphila* antennae has different receptors in different regions that react to cancer cell odors differently. When imaging all regions of the antennae, a difference map can be developed, which is highly sensitive to the odor changes of people with cancer [13].

A rat taste neuron was used both to study taste cell responses as well as to develop a sensor for sour taste [14]. The set-up is similar to a patch-clamp system (see Section 1.6), but instead of rupturing the membrane to measure the transmembrane potential, the potential change due to the change in ion concentrations is measured on the outside. The sour taste neuron was grown on a silicon chip, in this case a light-addressable potentiometric sensor (LAPS) chip [14] (Figure 4.4). This system was stable for up to 30 minutes. Different types of signals were measured; the signal had to be analyzed either in the time or the frequency domain for optimal results, depending on the type of signal. Sour signals between pH 2 and 4 could be detected. When several cells with all types of taste receptors were used as the sensor, time histograms and interspike histograms combined could be used to identify specific stimuli (e. g., sweetness) [15].

A similar system was used for the detection of smell [16]. A mixture of different rat olfactory neurons was grown on a LAPS chip until maturity (3 days), and then their signals were recorded. A significant difference between inhibitory and stimulatory signals could be detected. To make the sensor specific, the neurons were genetically engineered to express a specific odorant receptor, ODR-10 [17]. The result was a sensor that was specific for diacetyl, the natural ligand of ODR-10. The sensor also exhibited different firing frequencies for different concentrations of ligand. The detection range was 0.1 mM to 100 mM, but with limited reproducibility. To make each different cell separately addressable, each cell was immobilized onto a microelectrode of a microelectrode array (MEA) chip via a DNA strand [18]. With improved sensitivity and reproducibility, this system could result in a nanosensor that can detect more complex smells for some time.

All cells react to chemical stimuli (food, hormones, signal transduction molecules), and thus can be used as chemical sensors. One example is a sensor based on endothelial cells detecting the signal transduction molecule nitric oxide (NO) [19] (Figure 4.5). The cells were immobilized onto graphene using the RGD peptide (RGD stands for arginine, glycine, and aspartic acid and is a common part in cellular recognition) and kept alive

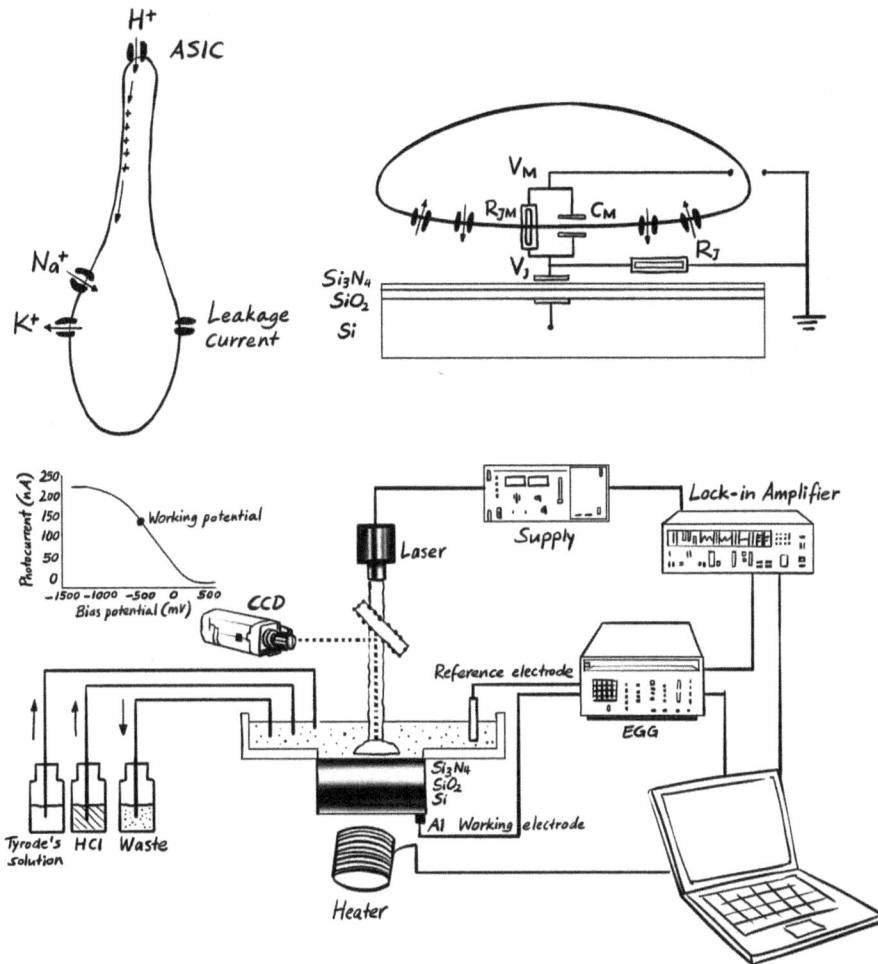

Figure 4.4: A rat taste neuron is grown on a silicon chip and used as a chemical sensor (adapted from [14]).

with a cell culture medium. NO was detected using double potential step chronoamperometry; the specificity was excellent even while being in solution. The detection limit was 25 nM and 80 nM of NO for a biofilm and separate cultured cells, respectively.

Instead of using cells, receptors and ion channels can be used directly as the sensing element. Here, two examples with different transducer elements will be highlighted. In one case, the goal was to detect two alcohols, 3-methyl-1-butanol and 1-hexanol, since salmonella bacteria release them when infecting beef [20]. To that end, a peptide, part of an odorant-binding protein from Drosophila called LUSH, was used as the sensing/binding element. The peptide was deposited on a quartz crystal microbalance (QCM) (see Section 1.6) and allowed to self-assemble. The change of resonance frequency of the sensor with and without bound alcohol was measured.

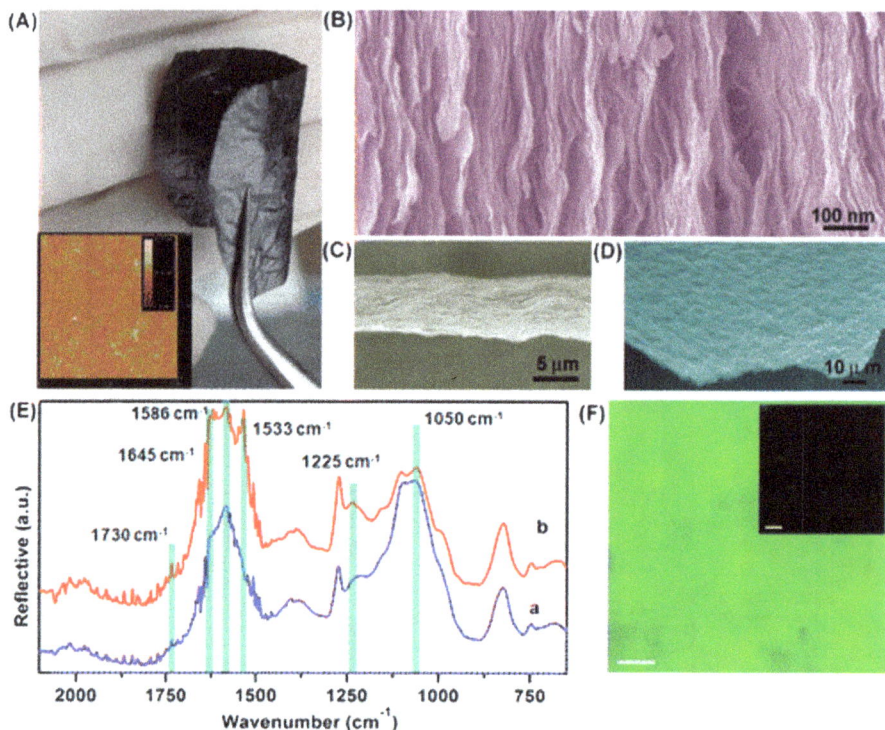

Figure 4.5: Cells are immobilized onto graphene and used as an NO sensor. NO is detected via chronoamperometry [19]. (A) Photograph of graphene biofilm. Side-view (B and C) and top-view (D) SEM images of the graphene biofilm. (E) IR spectra of (a) pyrenebutyric acid functionalized graphene film and (b) RGDpeptide covalently bonded graphene biofilm. (F) Fluorescence staining of graphene biofilm.

A common signal is shown in Figure 4.6 [20]. A concentration of 10 ppm of the alcohols (and possibly less) could be identified in about 20 s. Different sensors gave repeatable results within an acceptable error. When measuring mixtures of alcohols, the different alcohols could not be identified but the mixtures could still be measured. So, the sensor is not specific, but could be used in combination with specific sensors to increase the sensitivity of detecting rotting meat.

In another example, mouse G-protein-coupled receptors (GPCR) were used [21]. Three different receptors were coupled to carbon nanotube resistors. Since GPCRs are membrane proteins, they need to be stabilized inside a membrane. In this case, a "nanodisc" was used (Figure 4.7) (in nanodiscs, a lipid bilayer is surrounded by membrane scaffolding proteins that stabilize a small area of lipid bilayer). A change in current is measured when an odorant binds the receptor.

These devices are stable for at least a month. There is a significant drift in the response baseline, but that can be accounted for if one records the percent change from the baseline current (% $\Delta I/I$). Since the variability between different sensors is also a change

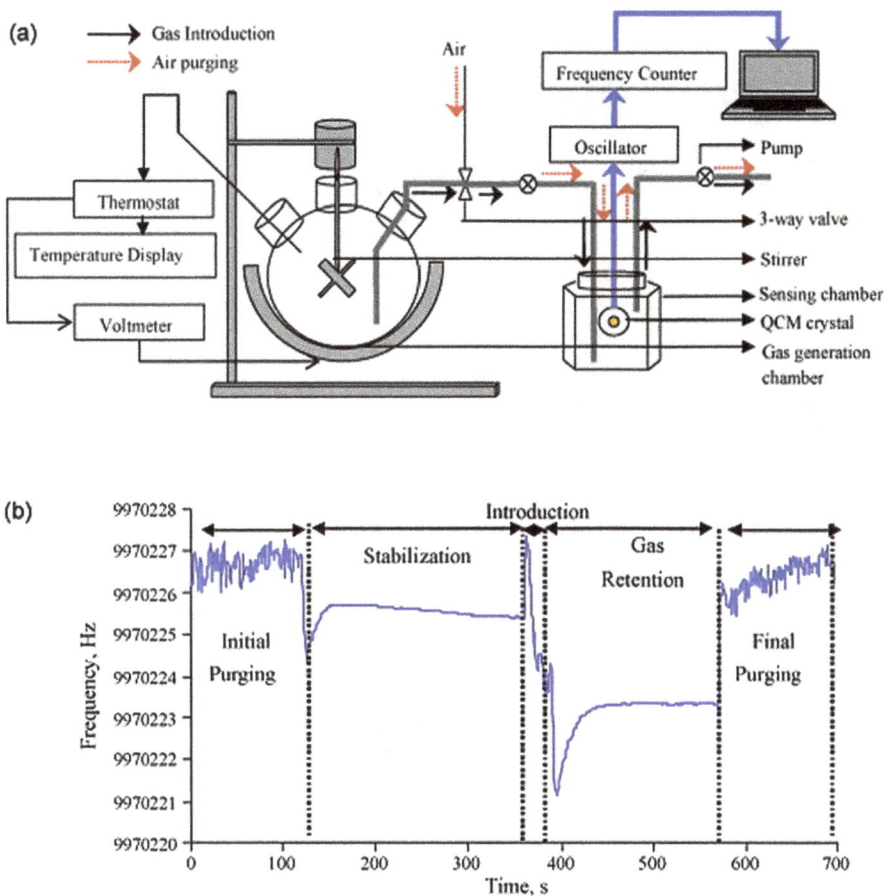

Figure 4.6: An odorant-binding protein peptide from Drosophila is deposited onto a quartz crystal microbalance to create a sensor for rotting meat smells (adapted from [20]). (a) Set-up of the biosensor; (b) signal of the sensor with alcohol present.

in baseline, this calculation makes different sensors comparable and reproducible as well. Each of the different GPCR sensors reacts differently to different chemicals, each in a concentration-dependent manner. Therefore, this sensor could in the future be combined with combinatorial identification capabilities, chemicals could thereby be identified out of mixtures.

That is already being attempted in the following example: The recognition element in that sensor is the sweet receptor binding domain, stabilized by a membrane that is immobilized on a gold electrode [22]. The transducer is a carbon nanotube field effect transistor. Binding of different sugars to the receptor domain changes the transistor current output, which can be analyzed. This sensor was independently verified by fluorescence quenching as well. This sensor was able to measure "sweet" in solution mixtures

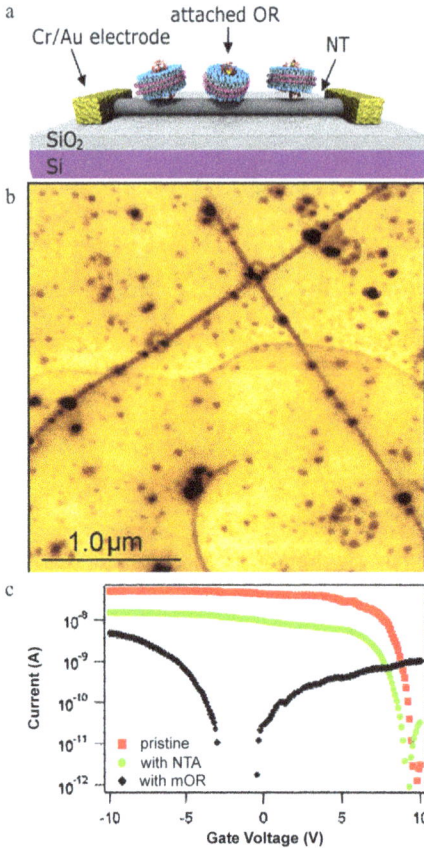

Figure 4.7: G-protein-coupled receptors (GPCR) are stabilized inside a membrane and attached to carbon nanotube resistors to create an artificial nose. Odorant-binding is detected via a change in current [21]. (a) Schematic of nanodisc sensor. (b) AFM image showing the attachment of nanodiscs to nanotubes. (c) IV curves of the same nanotube device as-fabricated (red), after functionalization (green), and after incubation in a solution of receptor protein micelles (black).

in femtomolar concentrations. It demonstrated one weird effect, though: chamomile enhances specifically the sucrose signal, even though chamomile itself does not bind to the sweet receptor binding domain [22].

So far, actual insect and mammalian cells and receptor molecules were used as the sensing element and connected to a transducing element, and thus detected. But how could one mimic the function of smell and taste detection and identification (i. e., the detection and identification of chemicals in gases and liquids, respectively) with nonbiological sensor molecules?

4.3 Biomimetic Chemical Sensors

To mimic smell and taste, several functions have to be incorporated: The chemical has to be taken up from water (taste) or air (smell), the molecule has to bind to a sensing molecule, and then the binding has to effect a change in a signal. The signal is then transduced or transferred to an analysis that makes sense of the signal. Not all of the functions are incorporated into all biomimetic sensors, but there are a large variety of different approaches for "artificial noses" and "artificial tongues" that have been reported. In this chapter, only a small selection will be described.

A common setup is shown in Figure 4.8 [23]. Some sensing molecule binds a specific compound or a specific type of compound with different strengths. The binding results in a changed signal. The signal is compared to a calibration and, in the case of multiple compounds, the binding molecule is identified through pattern recognition.

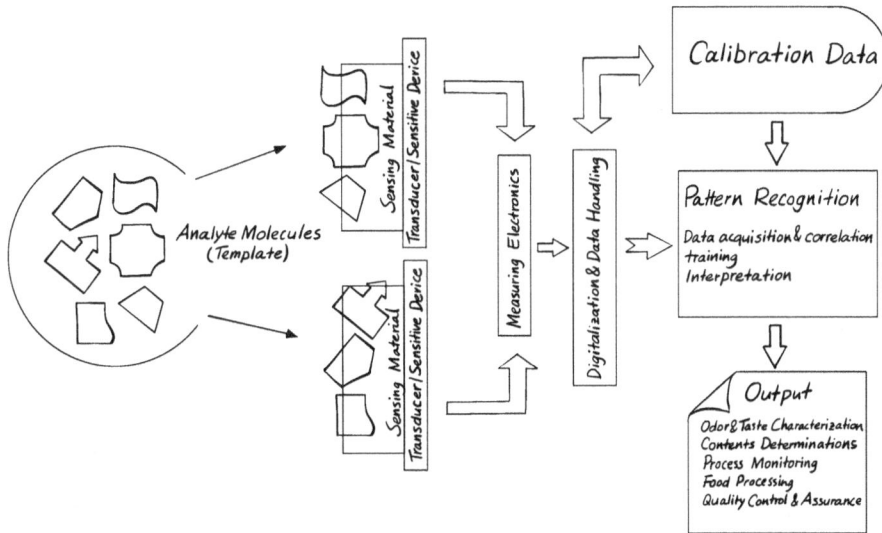

Figure 4.8: Common set-up for "artificial noses" and "artificial tongues" (adapted from [23]).

In the case of human smell and taste, the sample air or water is actively transported past the gas sensors in the nose (by breathing in air) and the aqueous sensors in the tongue (by swallowing the sample). Only a few of the reported sensors have tried to mimic this active obtainment of the molecule. In a microbead chemical sensor, capillary forces actively pull the sample into the microfluidic sensor [24]. In another example, channels were created that included the binding molecules in their inside wall [25, 26]. This method additionally results in size control—particles larger than the pore size were excluded from the sensor.

Mucin has also been used to recreate the active "catching" of the molecule [27]. An "E-tongue" was developed to identify bitter and astringent tastes only. It consists of and elastic hydrogel from a copolymer of acrylamide and acrylate mixed with chitosan. Mucin strongly binds the model substances for bitter taste (quinine sulfate) and astringent taste (tannic acid), but only weakly the model substances for sour taste (tartaric acid), and not at all the model substance for sweet taste (sucrose), salty taste (NaCl), and umami taste (monosodium glutamate). When measuring the voltammograms in different solutions, the sensor can determine different concentrations of bitter and astringent tastes in solution but is less effective when additional tastes are added to the mixture [27].

There is a large variety of different sensing molecules for different applications. A common method is a calorimetric sensor array using metalloporphyrins, solvatochromic dyes, or pH-sensitive dyes, which have been used for artificial noses [24, 28–30] and artificial tongues [31, 32]. Other optical sensors are based on fluorescence [33, 38] or luminescence [33]. Light sources, when needed, can be LEDs of different colors. Detectors are often photomultiplier tubes or CCD cameras [24, 32, 33]. An example for such a sensor is a calorimetric sensor developed for identifying different types of green tea [30] (Figure 4.9). The 36 sensor elements were different porphyrins that each reacted with a slightly different color to the different tea samples. A CCD camera took a picture before and after the sample and the difference between the two signals was identified in a difference map for all sensing elements, which was used to identify the tea sample.

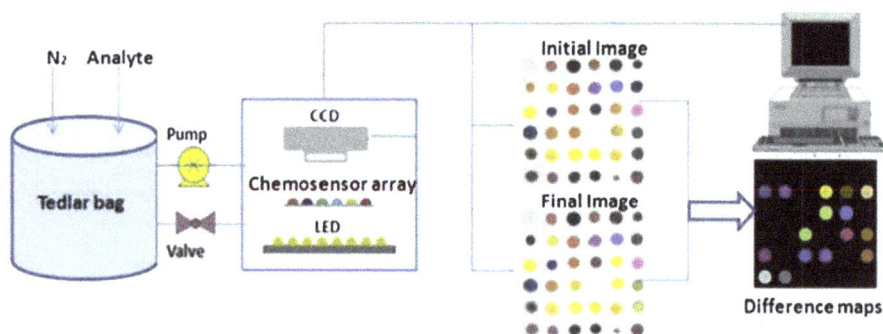

Figure 4.9: Calorimetric sensor with 36 sensor elements. Compounds are identified via pattern recognition of difference maps [30].

A similar sensor is used to identify gut bacteria (Figure 4.10) [34]. Twelve different fluorescence monomers in a PEG-block-PLL polymer are used as the sensing molecules. The fluorescence spectra are measured, and the difference pattern is analyzed by artificial intelligence for identifying different amounts and types of gut bacteria. The same sensor is used to identify differentiated stem cells [35].

An optical sensor is also used to measure glyphosphate concentration in solutions [36]. The sensing molecules are immobilized enzymes that have glyphosphate as their substrate. These enzymes are combined with competing glyphosate-functionalized poly(ethylene glycol) colloidal probes. Reflection interference contrast microscopy was used to measure interactions with a sensitivity down to 100 pM. This is a relatively low-

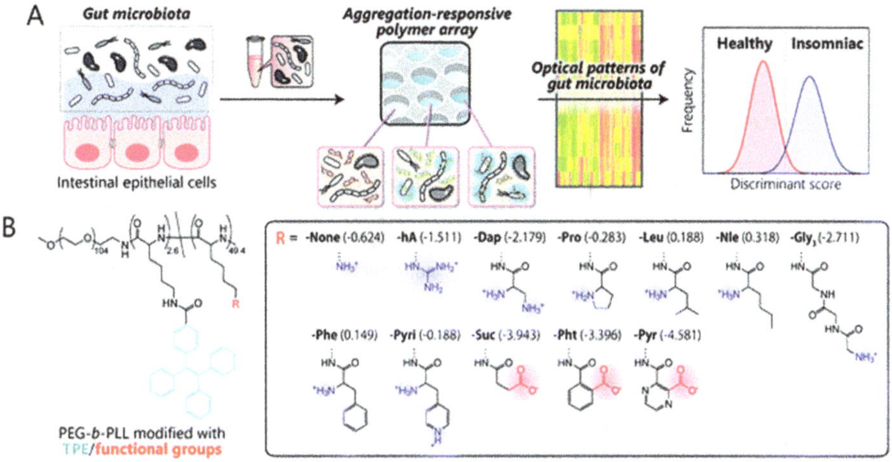

Figure 4.10: Twelve different fluorescence monomers in a PEG-block-PLL polymer bind gut bacteria differently. The fluorescence spectra are measured and analyzed by AI to identify different amounts and types of gut bacteria [34].

cost sensor that could eventually be used to develop an onsite sensing technology for food and water [36].

Instead of using enzymes as the sensing molecule, one can use an "enzyme mimic": imprinted polymers. Imprinted polymers are polymers synthesized around a template, which is then washed out, leaving a site specific for binding that template. In this example, a sensor for mycotoxin was developed using an imprinted polymer [37]. Mycotoxin is a toxin secreted from a fungus that attacks grain.

Mycotoxin fluoresces when irradiated with UV light, and a phone app was used to determine the amount of fluorescence. The best detection limit achieved was 20 ng/mL. Unfortunately, one needs to first prepare a grain extract to be able to use it, but this is a big step toward a cheap, effective mycotoxin sensor for grain and food.

Besides optical sensing methods, gravimetric sensors [32], specifically the quartz-crystal microbalance (QCM) [39], have been used for artificial tongues and noses. Electrical sensors [32, 33], based on semiconductors, MOSFETs (Metal Oxide Semiconductor Field Effect Transistors), conductive polymers, piezoelectricity, or carbon nanotubes, as well as gas chromatography have also been used [23].

An example for a QCM sensors uses self-assembled N-Acetylglucosamine (GLcNAc) monolayers [39]. GLcNAc is only part of the natural ligand for flu viruses, but even so the sensor can identify three different serotypes of the flu virus (Figure 4.11).

Applications for artificial noses and tongues are quite varied. In the food industry, it is used for process control, freshness determination, and authentication [23, 40]. Environmental and industrial monitoring is a common use for artificial noses, ranging from local monitoring of pollutant gases such as NO_x, SO_x, and CO from car and power plant exhausts or monitoring odors from landfills ([32] and references therein) to the

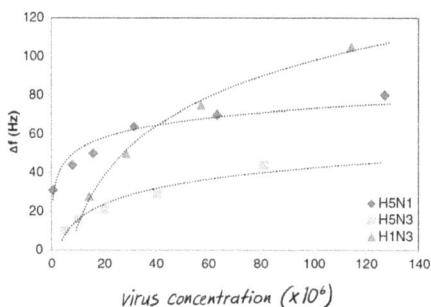

Figure 4.11: QCM sensor data differentiating between three serotypes of the flu virus. The sensor uses self-assembled N-Acetylglucosamine (GLcNAc) monolayers, which are part of the natural ligand for flu viruses as the sensor element [39].

detection of heavy metal ions [38]. The military is trying to use artificial noses to detect explosives and chemical warfare agents [32]. The other large and growing area of their application is in medical diagnostics [32]. These applications include the diagnostics of diseases from the patient's breath or the detection of different bacteria by looking at how they bind to different receptors.

4.4 Summary and the Bigger Picture

Of all the human senses, smell and taste have so far been most successfully mimicked, insofar as the terms of "artificial tongue" and "artificial nose" have been given to many nanosized chemical sensors regardless of whether they actually mimic the human processes or not. Chemical sensing methods are the furthest developed, often already combining multiple sensing elements with a transducer and an analysis element (generally a form of pattern recognition). Some of these sensors have already made it into industrial use. Even so, they still face a common problem: competing binding from unrelated chemical compounds reduces the specificity and reliability of analysis and compound identification. A preceding separation step could alleviate this problem.

Bibliography

[1]   Wong GT, Ruiz-Avila L, Ming D, Gannon KS, Margolskee RF. Cold Spring Harbor Symposia on Quantitative Biology. 1996;61:173.
[2]   Sharma A. Medical Biochemistry: Molecules to Disease. iBooks. 2016. https://itunes.apple.com/us/book/medical-biochemistry-molecules/id1119168051?mt=11.
[3]   Beherns M. In: Wiley Encyclopedia of Chemical Biology. 2007. p. 420–531.
[4]   Ramos Da Conceicao Neta E, Johanningsmeier SD, McFeeters RF. Journal of Food Science. 2007;72:R33–R38.
[5]   Behrens M, Meyerhof W, Hellfritsch C, Hofmann T. Angewandte Chemie, International Edition. International. 2011;50:2220–2242.
[6]   Margolskee RF. Pure and Applied Chemistry. 2002;74:1125–1133.
[7]   Bast WG, Albeanu DF. Nature Neuroscience. 2022;25(405):407.
[8]   DuBois GE. Proceedings of the National Academy of Sciences. 2004;101:13972–13973.

[9]   Hatt H. Praxis der Naturwissenschaften – Chemie in der Schule. 2009;58:6–11.

[10]   Bermúdez-Rattoni F. Nature Reviews Neuroscience. 2004;5:209–217.

[11]   Liu Y, Chen Q, Man Y, Wu W. Applied Mechanics and Materials. 2014;461:822–828.

[12]   Myrick AJ, Park KC, Hetling JR, Baker TC. Bioinspiration and Biomimitecs. 2008;3:046006.

[13]   Strauch M, Luedke A, Munch D, Laudes L, Galizia CG, Martinelli E, Lavra L, Paolesse R, Ulivieri A, Catini A, Capuano R, Di Natale C. Scientific Reports. 2014;4:3576.

[14]   Chen P, Liu XD, Wang B, Cheng G, Wang P. Sensors and Actuators B. 2009;139:576–583.

[15]   Chen P, Wang B, Cheng C, Wang P. Biosensors and Bioelectronics. 2009;25:228–233.

[16]   Wu C, Chen P, Yu H, Liu Q, Zong X, Cai H, Wang P. Biosensors and Bioelectronics. 2009;24:1498–1502.

[17]   Dua L, Wua C, Peng H, Zhao L, Huang L, Wang P. Biosensors and Bioelectronics. 2013;40:401–406.

[18]   Dua L, Zoua L, Wang Q, Zhao L, Huang L, Wang P, Wu C. Sensors and Actuators B. 2015;217:186–192.

[19]   Guo CX, Ng SR, Khoo SY, Zheng X, Chen P, Li CM. ACS Nano. 2012;6:6944–6951.

[20]   Sankaran S, Panigrahi S, Mallik S. Biosensors and Bioelectronics. 2011;26:3103–3109.

[21]   Goldsmith BR, Mitala JJJ, Josue J, Castro A, Lerner MB, Bayburt TH, Khamis SM, Jones RA, Brand JG, Sligar SG, Luetje CW, Gelperin A, Rhodes PA, Discher BM. JATC ACS Nano. 2011;5:5408–5416.

[22]   Jeong JY, Cha YK, Ahn SR, Shin J, Choi Y, Park TH, Hong S. ACS Applied Materials Interfaces. 2022;14:2478–2487.

[23]   Rehman A, Iqbal N, Lieberzeit PA, Dickert FL. Monatshefte Chemie. 2009;140:931–939.

[24]   Sohn YS, Goodey A, Anslyn EV, McDevitt JT, Shear JB, Neikirk DB. Biosensors and Bioelectronics. 2005;21:303–312.

[25]   Martınez-Manez R, Sancenon F, Biyikal M, Hecht M, Rurack K. Journal of Materials Chemistry. 2011;21:12588–12604.

[26]   Lee T, Lee HL, Tsai MH, Cheng SL, Lee SW, Huc JC, Chen LT. Biosensors and Bioelectronics. 2013;43:56–62.

[27]   Khan A, Ahmed A, Sun BY, Chen YC, Chuang WT, Chan YH, Gupta D, Wu PW, Lin HC. Biosensors and Bioelectronics. 2022;198:113811.

[28]   Janzen MC, Ponder JB, Bailey DP, Ingison CK, Suslick KS. Analytical Chemistry. 2006;78:3591–3600.

[29]   Rakow NA, Suslick KS. Nature. 2000;406:710.

[30]   Huo D, Wu Y, Yang M, Fa H, Luo X, Hou C. Food Chemistry. 2014;145:639–645.

[31]   Hou C, Dong J, Zhang G, Lei Y, Yang M, Zhang Y, Liu Z, Zhang S, Huo D. Biosensors and Bioelectronics. 2011;26:3981–3986.

[32]   Stitzel SE, Aernecke MJ, Walt DR. Annual Review of Biomedical Engineering. 2011;13:1–25.

[33]   Oh EH, Song HS, Park TH. Enzyme and Microbial Technology. 2011;48:427–437.

[34]   Tomita S, Kusada H, Kojima N, Ishihara S, Miyazaki K, Tamakice H, Kurita R. Chemical Science. 2022. https://doi.org/10.1039/d2sc00510g.

[35]   Tomita S, Ishihara S, Kurita R. Journal of Materials Chemistry B. 2022. https://doi.org/10.1039/d2tb00606e.

[36]   Rettke D, Doring D, Martin S, Venus T, Estrela Lopis I, Schmidt S, Ostermann K, Pompe T. Biosensors and Bioelectronics. 2020;165:112262.

[37]   Sergeyeva T, Yarynka D, Piletska E, Linnik R, Zaporozhets O, Brovko O, Piletsky S, El'skaya A. Talanta. 2019;201:204–210.

[38]   Xu W, Ren C, Teoh CL, Peng J, Gadre SH, Rhee HW, Lee CLK, Chang YT. Analytical Chemistry. 2014;86:8763–8769.

[39]   Wangchareansak T, Sangma C, Ngernmeesri P, Thitithanyanont A, Lieberzeit PA. Analytical and Bioanalytical Chemistry. 2013;405:6471–6478.

[40]   Peris M, Escuder-Gilabert L. Analytica Chimica Acta. 2009;638:1–15.

5 Hearing

5.1 Human Hearing on the Molecular Scale

All sensors, biological or technological, have several elements: the sensing element that senses the signal, the transducer that transfers the signal, and an amplification and/or analysis/reporting element that increases the signal and/or analyzes it. In the case of hearing, the energy of a soundwave (which could also be called the vibration of air) is turned into the vibration of the eardrum or tympany, which vibrates small bones to transfer the vibration into the liquid of the inner ear (Figure 5.1) [1, 2]. These bones are necessary because air and water have different properties, and if the vibration was transferred directly from air to water the majority of the sound energy would be lost. The ear drum is the connection between the outer and the middle ear. The oval window is the connection between the middle ear and the inner ear, which is another membrane transferring vibrations. The inner ear contains the cochlea, which contains three different compartments with different liquids. Two of these compartments are separated by the basilar membrane (Figure 5.2). On the surface of that membrane, the transduction from vibration to electrical signal takes place: Hair cells contain bundles of cilia or hairs of different lengths, and the top of each bundle is connected to a neighboring bundle via a helical protein that can stretch and relax (Figure 5.3). Vibration of the basilar membrane coming from the sound waves moves the hair bundles toward another membrane (tectorial membrane), which bends the larger bundles before it reaches the smaller bundles. That stretches the helical proteins at the line between bent and straight bundles. When the connection is stretched, it opens mechanoelectrical transduction (MET) ion channels on the top of the hair cells, transporting potassium ions into the cells. There is now an imbalance of charge across the membrane, i. e., the membrane is depolarized. Depolarization causes glutamate, a neurotransmitter, to be released into the surrounding area close to a cochlear nerve cell. The membrane of the nerve cell becomes depolarized, thus starting an action potential that transfers to the brain. A combination of brain cells then allows for the interpretation of the original signal (the sound waves), thus you now hear a certain tone, chord, or sound.

Even bending the hairs for only 1 nm creates a signal, and thus "hearing" [3]. The frequency of opening the ion channel is likely going to aid in hearing different tones. The exact ion channels involved have not been identified yet, but there are likely at least two different types.

Why is there such a long, arduous path for the signal, transporting the signal from air through liquid via several membranes? This is important because the majority of sounds are a mixture of wavelengths. The cleanest way to transport and identify their mixture is to separate out the different wavelengths so that each can be identified separately, and then put only back together into the original sound in the brain. The basilar membrane plays a role in this "mechanical filtering" [4]. The interaction of the hairs with the tectorial membrane is part of the amplification of the signal. This combination

https://doi.org/10.1515/9783110779196-005

EAR

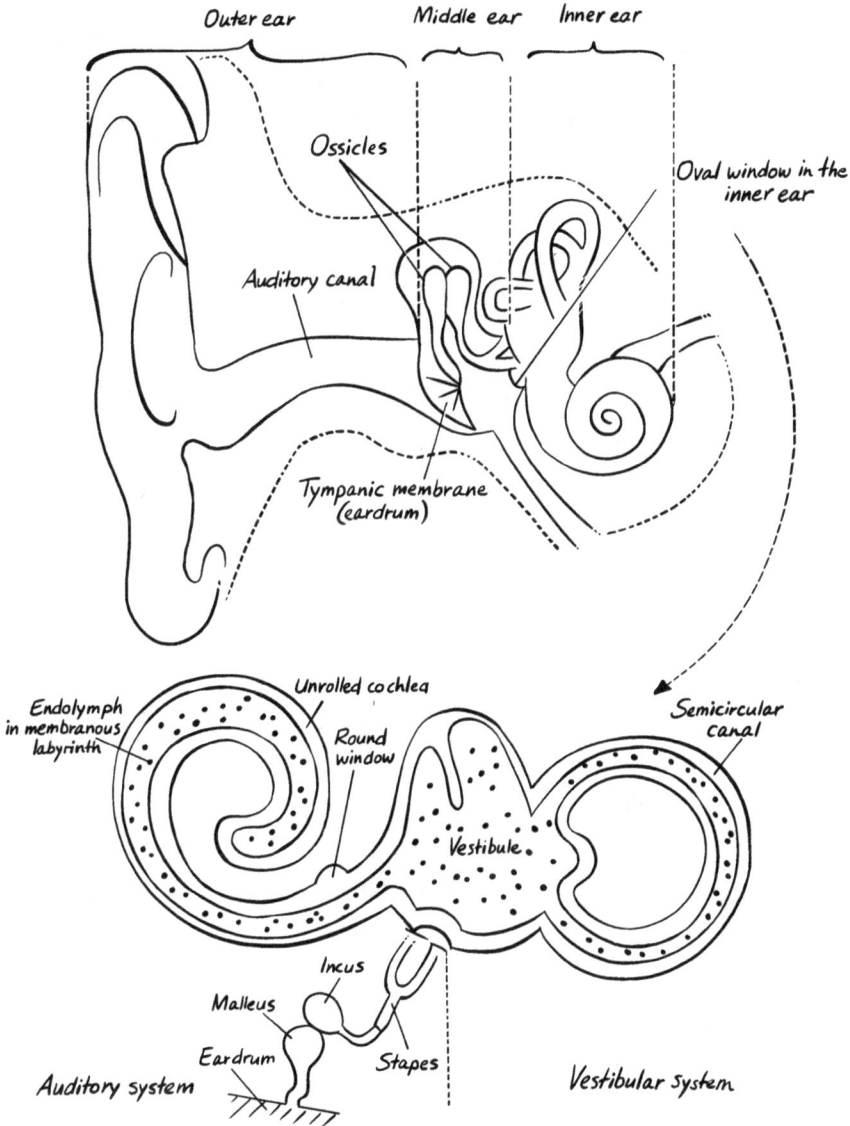

Figure 5.1: The structure of the human auditory system and how hearing works on the molecular scale (adapted from [1]).

of separating each wavelength and then amplifying them makes our hearing a lot more sensitive.

How can a lipid bilayer membrane perform such functions as separating out wavelengths and amplifying them? This process is not fully understood but it is based on the

COCHLEA

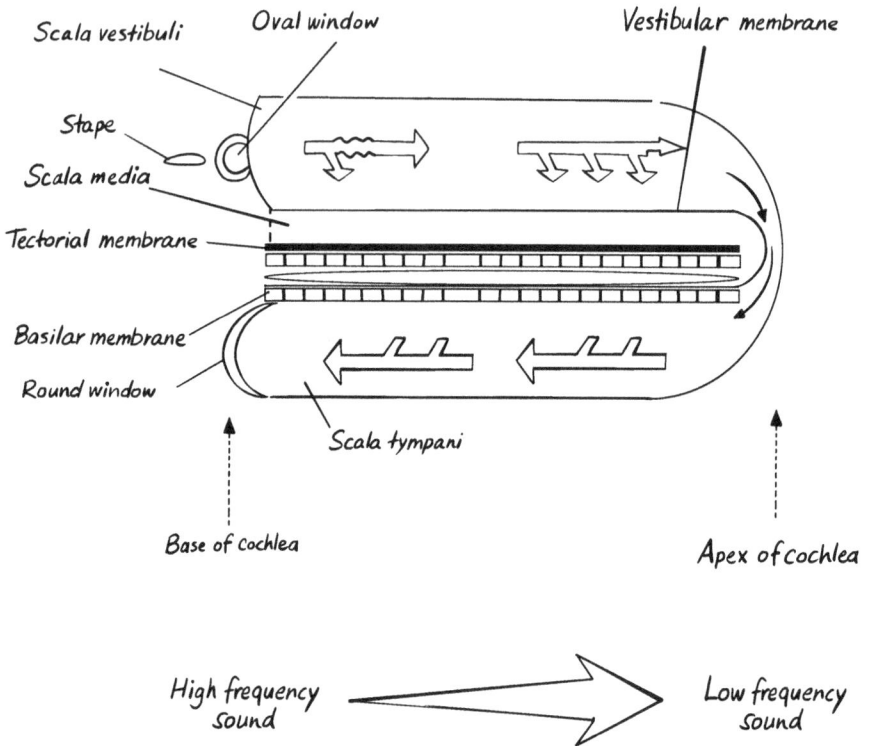

Figure 5.2: The inner ear with the basilar membrane.

membrane stiffness. The membrane can control the stiffness of its different sections by first modifying the composition of a section (i. e., more cholesterol generally stiffens a membrane) as well as anchoring it at specific points to the cytoskeleton at the inside of the cell and possibly the outside via special proteins [5]. Stiffer parts dampen vibrations, softer ones enhance them.

Let's summarize: sound waves in air are transferred into sound waves in liquids, which makes membranes vibrate. The long path of the vibrations results in the separation and amplification of the different wavelengths of the sound. The transduction of the vibration to the electrical signal occurs because stretching of a protein according to those vibrations mechanically opens an ion channel, which releases a neurotransmitter that starts the electrical transduction throughout the nerves and brain. Only the workup of the electrical signal puts all of the wavelengths together again and identifies the original signal as a specific sound. Can the molecules and methods of the human auditory system be used to make an artificial, molecular-sized auditory sensor with similar functions?

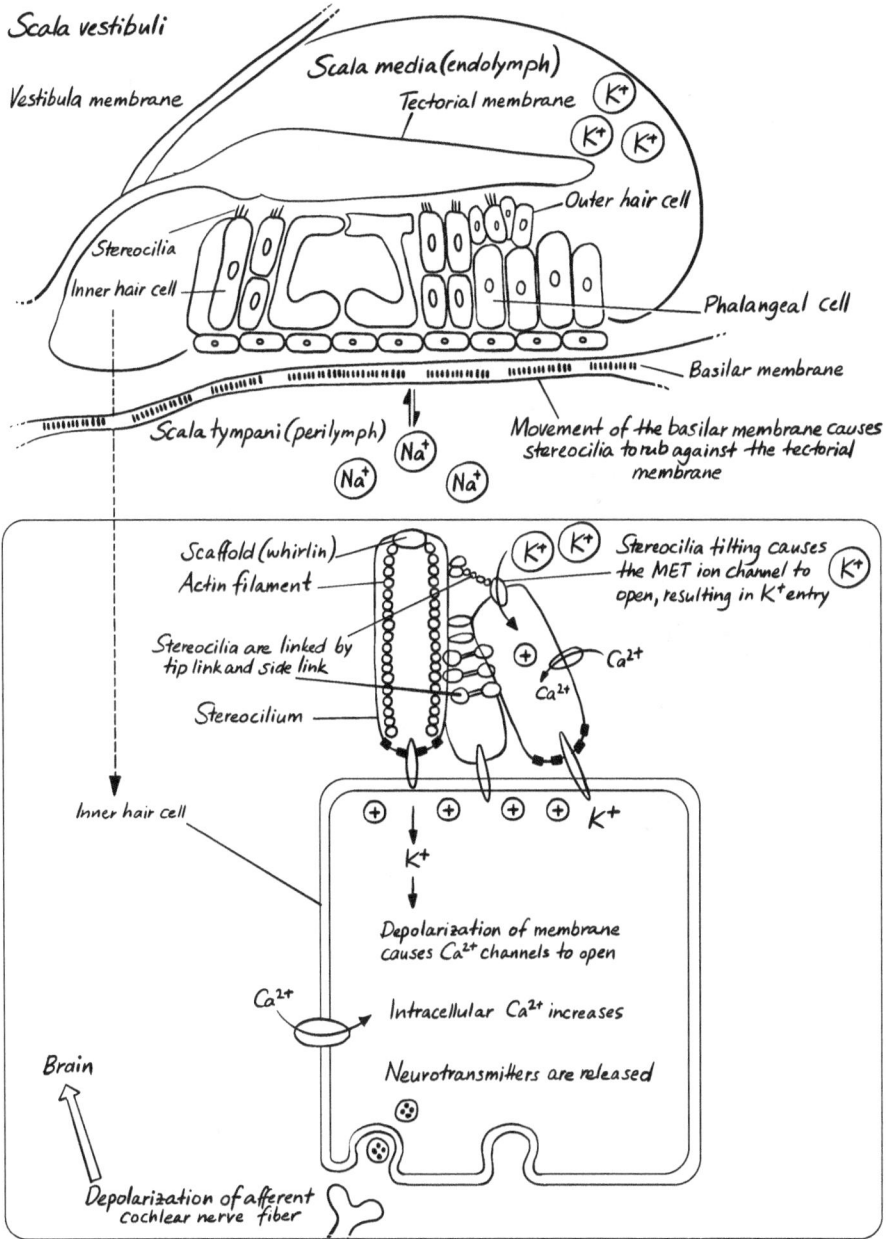

Figure 5.3: Vibration of the basilar membrane results in bending hair bundles, which opens ion channels on the top of the hair cells, which eventually starts an action potential.

5.2 Vibration Sensors Using Biological Cells, Molecules, and Methods

Receivers and microphones are a mature technology and have been well miniaturized, so that the need for nanovibration sensors using biological molecules and cells is small and only very few examples exist. The majority of these sensors are developed to study the human ear itself, specifically its hair cells.

One of the problems in studying hair cells is that it is necessary to stimulate and then monitor the hair cells at specific frequencies to understand their mechano-electrical transduction. Ideally, the force should be delivered instantaneously by a spring of a very precise stiffness (spring constant), and the effect of the stimulation should be measured by a rigid fiber. Such a system has been developed using nanomechanical force probes [6] (Figure 5.4). These nanomechanical force probes included integrated piezoresistive sensing and piezoelectric actuation. The spring constant was in the necessary range and the probe was capable of delivering mechanical stimuli with sub-10 µs rise times in water.

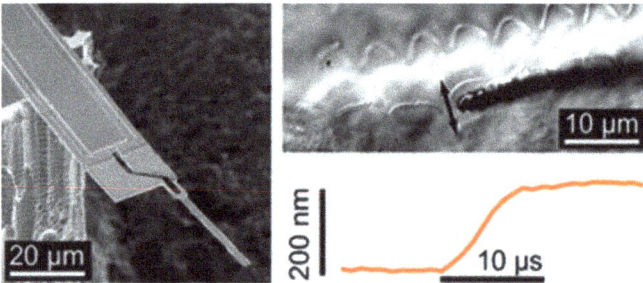

Figure 5.4: A nanomechanical force probe is used to measure the mechano-electrical transduction of the hair cells at specific frequencies [6].

Another approach to controlling the hair bundles of the hair cells is to use magnetic particles [7] (Figure 5.5). The control is supposed to be remote, reversible, and localized. This has been achieved by coating specific hair cells with cubic nanoparticles; each cell could then be stimulated to remotely open and close its MET ion channel at time scales that varied from a few seconds to 100 µs. The effect could be measured by measuring calcium ion flows via a fluorescent probe. This system was effective for studying individual hair cells as well as other systems with mechanosensitive ion channels.

In summary, no reports were found in which biological molecules or cells were used to create a biomimetic sound receiver. But with the research done to study the mechanosensitive-electrical transduction process of the ear it might be possible in the future to mimic the wavelength separation and amplification of the membranes, thus leading to more sensitive receivers on the nanoscale.

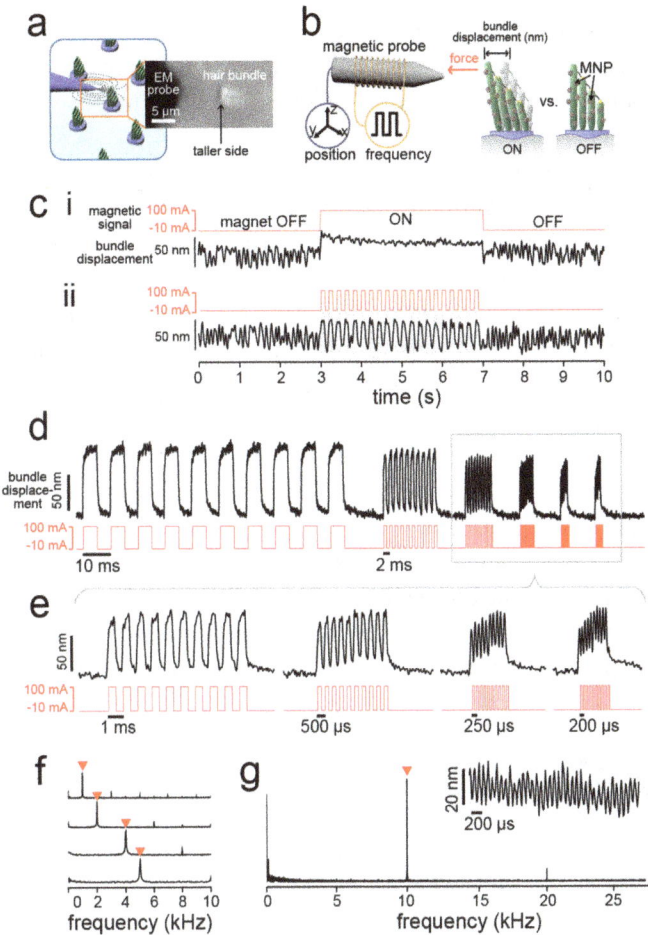

Figure 5.5: Magnetic nanoparticles are used to control the opening and closing of the MET ion channels connected to hair cells at time scales that varied from a few seconds to 100 μs (adapted from [7]). (a) Schematic drawing and microscope image of a magnetic probe activating a single hair bundle. (b) Schematic diagram of the EM probe stimulating a single hair bundle with magnetic particles. (c) Motion of a single hair bundle deflected by the EM probe. Black line: tracked bundle displacement, red line: current applied to the EM probe. Application of (i) static and (ii) 5 Hz square wave current. (d), (e) Varying frequency stimulation. (f) Fourier transform spectra of oscillation shown in (e). (g) FT spectrum of bundle motions at 10 000 Hz.

5.3 Biomimetic Acoustic Sensors

The most common biomimetic (nano) technology for hearing is the microphone. It contains a diaphragm, which mimics the eardrum of the oval window, which vibrates according to the sound in the environment. The mechanical movement of the diaphragm is then converted into an electrical signal. The diaphragm could be part of a plate capacitor, in that case the electrical signal would be caused by the difference in voltage due to the different distance [8]. Alternatively, a material with a permanent charge (an "electret" of "ferroelectric" material) could be used with the same effect [9]. These methods are commonly used in microphones.

On a smaller scale, vibration can also be captured by a piezoelectric crystal [10]. Piezoelectricity by definition is a transducer of vibration/pressure to electricity—these materials emit voltage when under pressure. In a piezoelectric microphone, the transducer is such a crystal.

Microphones can also be so small that they can be part of a microelectromechanical system (MEMS), a special type of computer chip [11]. In this case, the pressure-sensitive diaphragm is directly etched out of the silicon in the chip. Microphones made in this way are digital microphones.

Here, I would like to focus on biomimetic acoustic sensors that are either implantable, mimic the attributes hearing possesses additionally to microphones, or use the acoustic sensor for a different application.

As an example of implantable microphones, Lee's group is in the process of developing an artificial hair-cell microphone based on a flexible, piezoelectric film [12] (Figure 5.6). This system exhibits some of the frequency separation and selectivity of the basilar membrane. It was possible to align the distribution of vibration displacement

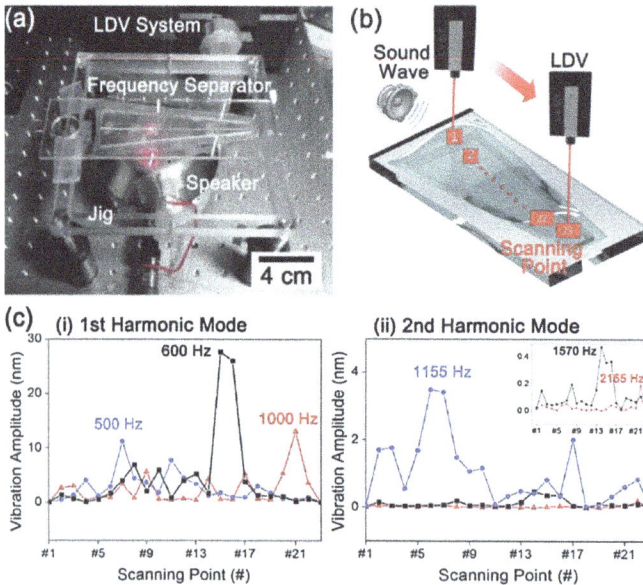

Figure 5.6: Artificial hair-cell microphone based on a flexible, piezoelectric film developed for implementation [12]. (a) Experiment setup for measuring the vibration amplitude of the flexible piezoelectric film in response to a sound. (b) The film is scanned to detect vibrations by using scanning points. (c) Peak of vibration over all scanning points in (i) first harmonic mode and (ii) second harmonic mode.

from the frequency separation closely with that of the piezoelectric signals. The same effect could also be produced with different-length beams on a field-effect transistor as part of a MEMS acoustic sensor [13].

It is useful to measure vibrations of a variety of wavelengths for an array of medical applications. For wearable sensors, they have to be made from flexible materials and be very sensitive. One example uses a network of multiwalled carbon nanotubes (CNT) in a suspended membrane (Figure 5.7) [14]. The cracks in the MWCNT layer increase flexibility and change the resistivity, increasing the sensitivity of the sensor (Figure 5.7:(a)i). The suspension of the membrane allows for larger vibrations and with thus increased sensitivity as well (Figure 5.7:(a)ii). This sensor can be used for a large range of frequencies, but also for the detection of different arm movement.

Vibration sensors can also be used to measure flow (Figure 5.8) [15]. In this example, a hair is incorporated into the bilayer interface, and electrodes are fed into each aqueous phase, resulting in a sensitive flow sensor.

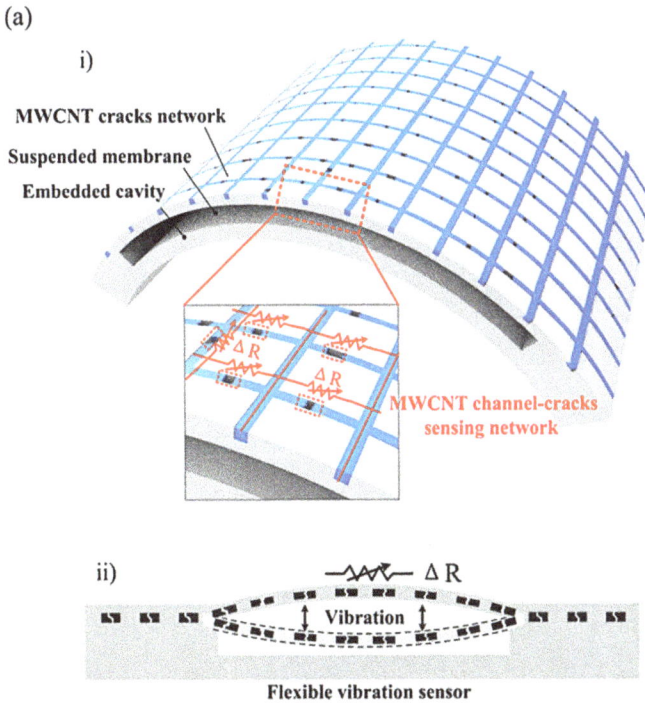

Figure 5.7: (a) Scheme of the layers and set up of the vibration sensor. (i) MVCNT layer with cracks in a suspended membrane (ii) vibration enhancement due to the suspension of the membrane. (b) Sensor output for a large frequency range. (c) Sensor outputs for arm movement (top) and different sounds (bottom). With permission from [14].

(b)

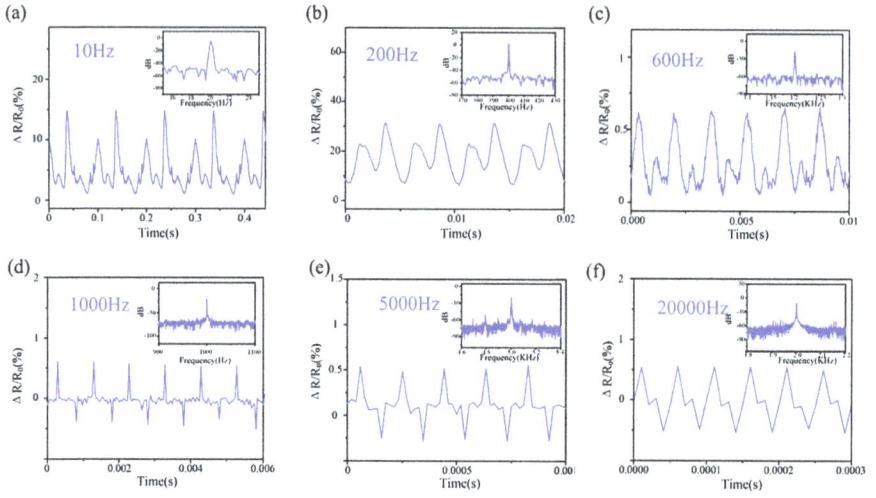

(c)

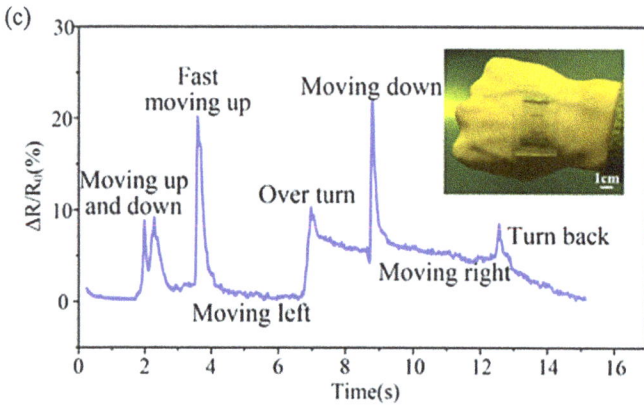

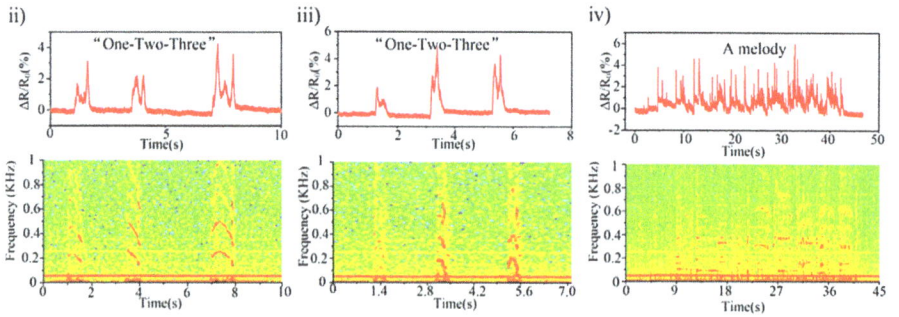

Figure 5.7: (continued)

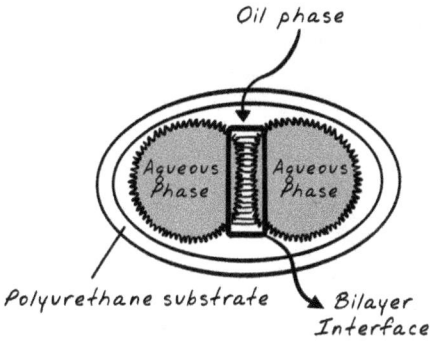

Figure 5.8: A hair is incorporated into the bilayer interface, and electrodes are fed into each aqueous phase, resulting in a sensitive flow sensor.

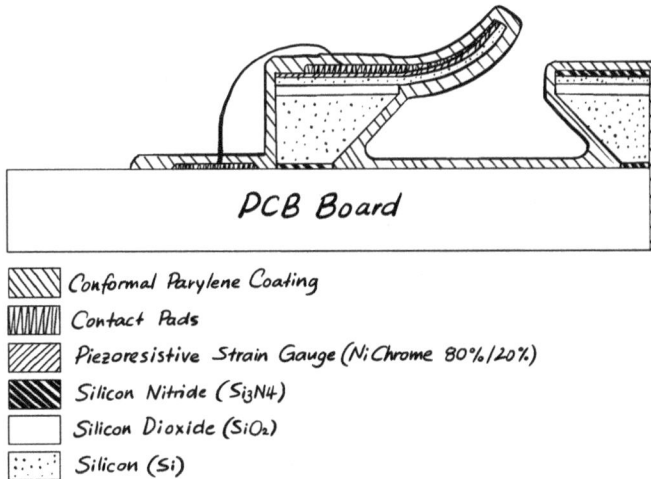

Figure 5.9: Artificial hair cells used as flow sensors (adapted from [16]).

Another flow sensor uses a silicon beam accompanied by a piezoelectric strain sensor [16] (Figure 5.9). The sensor was mechanically stable even at a high flow velocity. This artificial hair cell was also able to sense the flow direction and low frequencies in water.

In a different system, vibrations are not measured, but their energy is harvested to create a "battery" for implanted devices such as pacemakers and deep brain stimulation devices [17]. Conventional batteries are problematic for these devices because they are bulky and have to be replaced regularly. It would be a lot more effective if an energy source could be harvested that is already part of the body and could be used to power these devices. To develop such a system, a film of a piezoelectric material on a flexible substrate could be used, which allows the piezoelectric material to react to any bending by generating an electric signal. Vibrations are sufficient to generate a signal, which

(a)

i) PMN-PT Thin Film on Bulk Wafer   iii) PMN-PT Thin Film on Flexible Substrate

ii) Exfoliation of MIM Structure from Wafer   iv) Direct Pacemaking

(b)   **(c)**

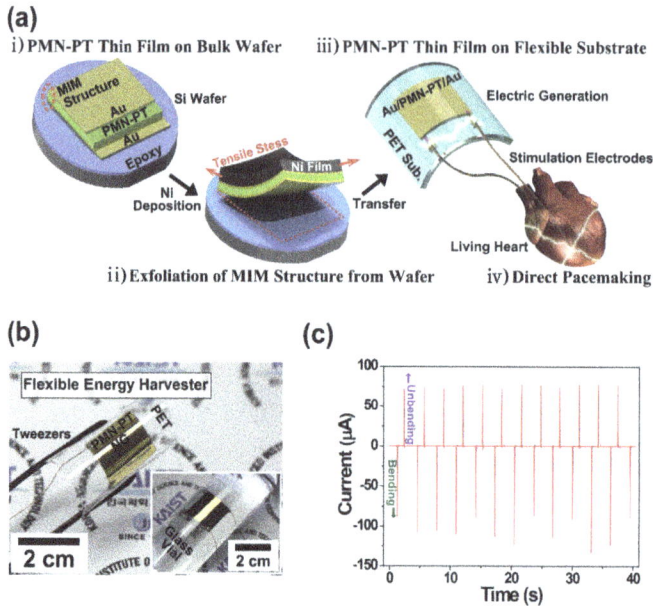

Figure 5.10: A piezoelectric material on a flexible substrate is used as an energy source for implanted devices [18]. (a) Schematic illustration of the fabrication process and biomedical application of flexible PMN-PT piezoelectric energy harvester. (b) Cross-sectional SEM image of the PMN-PT thin film on a PET substrate. The inset shows an XRD pattern of PMN-PT thin film. (c) Raman spectrum obtained from PMN-PT thin film. The indexed sharp spectra agrees well with the typical features of perovskite PMN-PT.

makes it also effective as a vibration sensor. But the film can also be used as an energy source for implanted devices (Figure 5.10) ([17] and references therein).

5.4 Summary and the Bigger Picture

Our acoustic sensor, the ear, is actually rather complicated: vibrations in air are converted into mechanical vibrations in a membrane, into vibrations in water, back into mechanical vibrations of a membrane that then are transduced into electricity via the bending of fiber that mechanically opens an ion channel. Why is the ear so complicated, when a microphone can record sound much more easily and still produce excellent sound quality? Basically, the sensitivity of the ear is important because any sound is a mixture, and the basilar membrane is able to separate out the wavelengths and amplify and detect each of them separately. The wavelengths are only mixed together again in the brain to create the original sound.

There is research currently being carried out into biomimetic hair cells to try to copy these additional features on several length scales, but no one has yet achieved a microphone with the efficiency of the basilar membrane. Additionally, vibration sensors can be used as flow sensors and energy harvesting devices.

Bibliography

[1] Sharma A. Medical Biochemistry: Molecules to Disease. iBooks. 2016. https://itunes.apple.com/us/book/medical-biochemistry-molecules/id1119168051?mt=11.

[2] Fettiplace F, Kim KX. Physiological Reviews. 2014;94:951–986.

[3] Katta S, Krieg M, Goodman MB. Annual Review of Cell and Developmental Biology. 2015;31:347–371.

[4] Kazmierczak P, Müller U. Trends in Neurosciences. 2012;35:221.

[5] Anishkina A, Kung C. Proceedings of the National Academy of Sciences. 2013;110:4886–4892.

[6] Doll JC, Peng AW, Ricci AJ, Pruitt BL. Nano Letters. 2012;12:6107–6111.

[7] Lee JH, Kim JW, Levy M, Kao A, Noh SH, Bozovic D, Cheon J. ACS Nano. 2014;8:6590–6598.

[8] Huffman L. Stokowski.org; http://www.stokowski.org/Development_of_Electrical_Recording.htm.

[9] Sessler GM, West JE. The Journal of the Acoustical Society of America. 1962;34:1787–1788.

[10] Lee WS, Lee SS. Sensors and Actuators A. 2008;144:367–373.

[11] Johnson RC. In: EETimes, 2014.

[12] Lee HS, Chung J, Hwang GT, Jeong GT, Jung Y, Kwak JH, Kang H, Byun M, Kim WD, Hur S, Oh SH, Lee KJ. Advanced Functional Materials. 2014;24:6914–6921.

[13] Mastropaolo E, Latif R, Koickal T, Hamilton A, Cheung R, Newton M, Smith L. Journal of Vacuum Science and Technology B. 2012;30:06FD01-01–06FD01-07.

[14] Chen X, Zeng Q, Shao J, Li S, Li X, Tian H, Liu G, Nie B, Luo Y. ACS Applied Materials Interfaces. 2021;13:34637–34647.

[15] Pinto PA, Garrison G, Leo DJ, Sarles SA. In: Proceedings of SPIE, Bioinspiration, Biomimetics, and Bioreplication, Vol 8339, 2012;833907. https://doi.org/10.1117/12.915198.

[16] Qualtieri A, Rizzi R, Epifani G, Ernits A, Kruusma M, De Vittorio M. Microelectronic Engineering. 2012;98:516–519.

[17] Hwang GT, Byun M, Jeong CK, Lee KJ. Advanced Healthcare Materials. 2015;4:646–658.

[18] Hwang GT, Park H, Lee JH, Oh S, Park HI, Byun M, Park H, Ahn G, Jeong CK, No K, Kwon H, Lee SG, Joung B, Lee KJ. Advanced Materials. 2014;26:4880–4887.

6 Skin, The Body's Largest Organ

6.1 Human Skin on the Molecular Scale

All sensors, biological or technological, contain several elements: the sensing element that senses the signal, the transducer that transfers the signal, and an amplification and/or analysis/reporting element that increases the signal and/or analyzes it. Skin is the largest and most complex organ in the body. It has a variety of sensor functions, each with a transducer that transfers and possibly amplifies the signal. These signals are then analyzed and identified in the brain.

The main sensors in skin are touch (mechanical forces), and heat/cold (temperature). Additionally, there are sensors for chemical stimuli. For all sensors, there are also pain receptors that might be the same or a different set of receptors. Both are connected to automatic actions of the body: avoidance, in the case of extreme heat or cold and chemicals; but also, the active temperature and fluid/electrolyte regulation of the body that skin is mostly responsible for—in the case of pressure, whatever action needed to release the pressure/avoid the pain. All of these actions are part of the skin's response to the outside environment, which skin is also responsible for, since it is closest to the outside environment.

One of the biggest other functions of skin, besides sensing and body regulation, is protection. Skin contains the organs so that they can function under precise conditions, called homeostasis. At the same time, skin protects from injury (including temperature and pressure), but it also protects from bacteria and viruses that could lead to disease. When injury does occur (not everything can be prevented), skin starts a self-healing process to contain and then heal the injury. Skin also has additional functions such as Vitamin D production.

All of these functions are contained in a relatively thin, two-layer membrane [1] (Figure 6.1). The structure is rather complex, but optimized for the combination of all the functions. This chapter will talk about the molecular structure of the senses in the skin first, then describe briefly the body regulation system in skin, the healing process, and the structure of skin that makes all of these functions possible.

Temperature is sensed by afferent neurons in the peripheral nervous system, i. e., the neurons that bring sensory information from the outside (skin) to the brain and spinal cord. There is only limited information on the neurons that sense temperature. It was originally assumed that the nerve endings for temperature sensing had to be different, but research shows that the difference mostly lies in the type of ion channel [2]. One type of ion channel that was identified and researched is the family of transient receptor potential ion channels. First found in the drosophila fly, by now four channels for heat and two for cold were found in humans (Figure 6.2). As the graph suggests, not only does each receptor have a specific temperature range it reacts to, each also reacts to specific chemicals. The chemicals are generally the ones that leave either a cooling or a burning sensation on the skin, which makes sense since they are primarily temperature

https://doi.org/10.1515/9783110779196-006

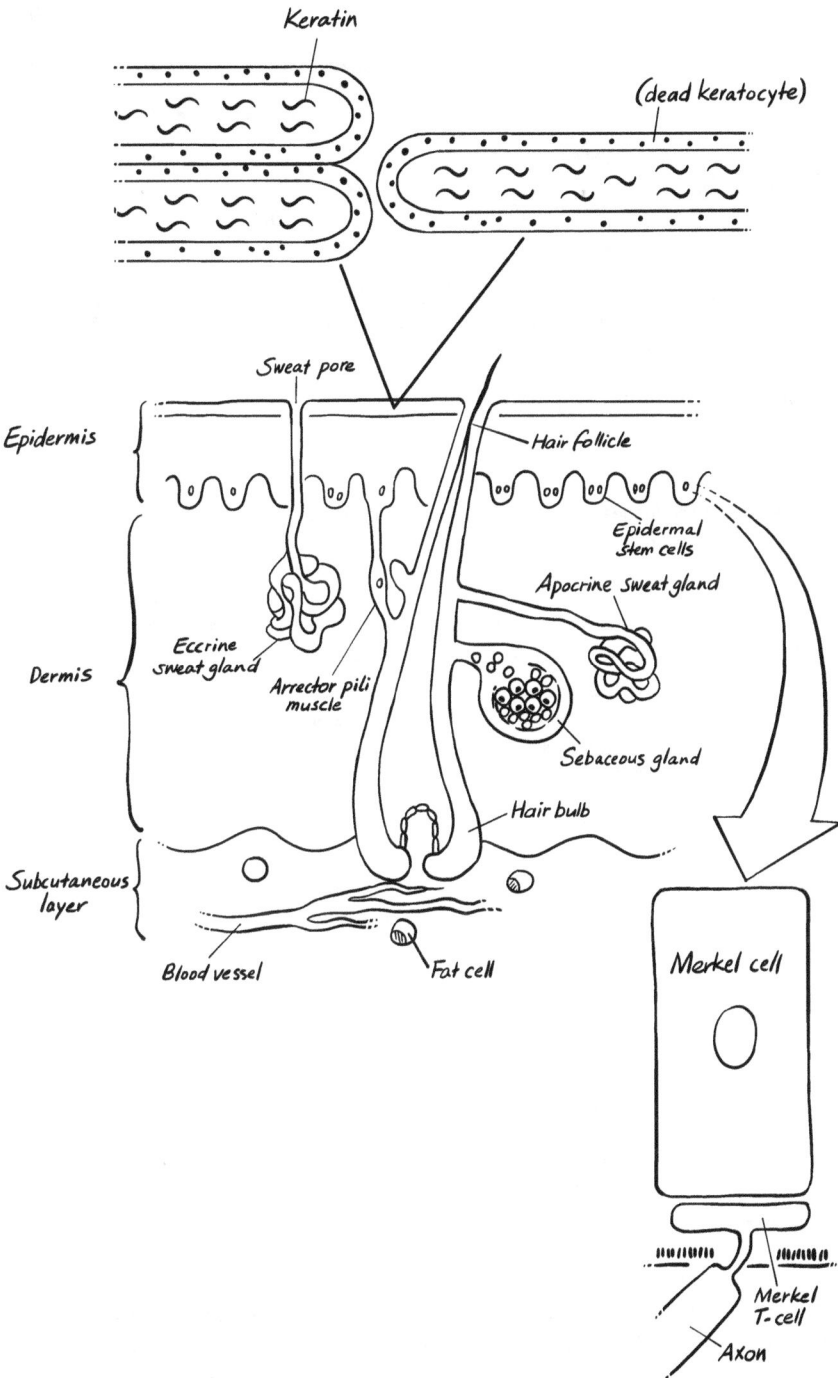

Figure 6.1: The structure of skin and the molecular function of its senses (adapted from [1]).

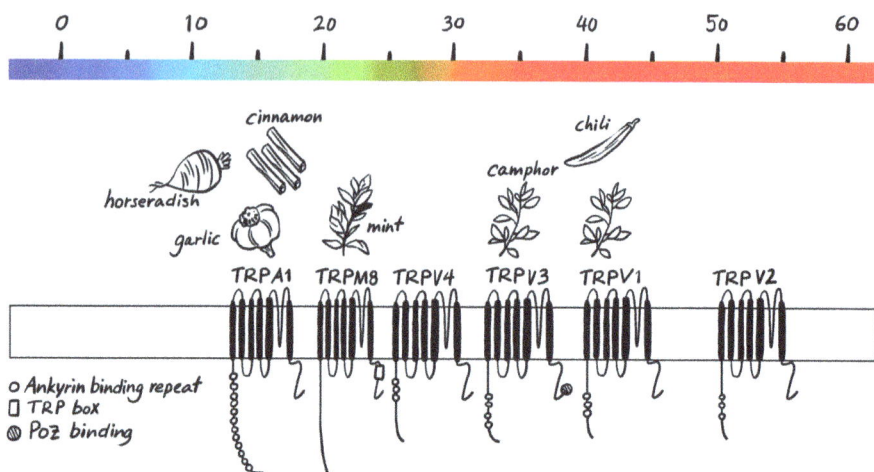

Figure 6.2: Four channels for heat and two for cold were found in the nerve endings for temperature sensing in the skin (adapted from [2]).

Table 6.1: Each receptor reacts to specific temperature ranges and specific chemicals. The channels that react to the chemicals are the noci-receptors of the series, i. e., the pain receptors [2].

Channel	Temperature Sensitivity	Common Nonthermal Agonists	Tissue Distribution
TRPV1	≥ 42 °C	Capsaicin, acidic pH, allicin, camphor	PNS, brain, spinal cord, skin, tongue, bladder
TRPV2	≥ 52 °C	Growth factors	PNS, brain, spinal cord
TRPV3	≥ 33 °C	camphor	PNS, skin
TRPV4	27 °C–42 °C	Hypotonic	PNS, brain, skin, kidney, inner ear, liver, trachea, heart, hypothalamus, fat
TRPM8	≤ 25 °C	Menthol, eucalyptol	PNS, prostate
TRPA1	≤ 17 °C	Cinnamaldehyde, mustard oil, allicin	PNS, hair cells
TREK-1	cold	Membrane stretch, intracellular pH	PNS, brain
P2X3	warmth	ATP	PNS
BNCI, ASIC	cold		PNS

sensors. The channels that react to the chemicals are the noci-receptors of the series, i. e., the pain receptors (Table 6.1). So "pain" is basically sensed as the extreme of a normal sensation, such as heat or cold, and thus is sensed by the same receptors using the same principles. Generally, the ion channel opens at a specific temperature. The mechanism by which the channels open and close is still under debate, but the thermodynamics of the three-dimensional channel structure seems to be at least part of the cause [2]. Most of these channels are also voltage dependent, and the voltage does change the temperature range somewhat. There are additionally some chemicals that allosterically regulate temperature range and/or voltage dependency. Even though each channel has its own temperature range, the actual sensation of heat can usually only be explained by the combination of several channels.

The ion channels that react to noxious (painful) temperatures seem to also be the ones that react to noxious chemicals, such as capsaicin from hot chili peppers [3] or acid in concentrations that cause burns [2] (Table 6.1). At the same time, each of these channels are located not only in skin, but also, e. g., in the tongue, some organs, and the central nervous system. Inside of organs, the channels seem to be connected with the inflammation response.

As soon as the channel is activated, the signal is either transferred directly to another neuron, and thus to the brain via a synapse, or transferred indirectly via a chemical release of the skin cell to activate another neuron that then has a synaptic connection to other neurons and the brain [2]. Skin cells can produce cytokines, neural growth factors, and neurotransmitters that all could function as an indirect signal. One path for the signals to the brain includes the limbic system, which generates an emotional response to pain [4].

Some of the channels seem also be part of the temperature regulation of the body as a result of sensing the temperature [2, 5]. At least locally, some of the channels initiate blood vessel dilation or restriction, which regulates temperature in that area of skin. The temperature control cycle is then connected to other homeostasis cycles, such as fat cycles in the brown adipose tissue that is used to internally create heat [6]. The details of most of these processes are not yet understood. Questions that remain are the specific gating mechanism of these channels, how they are regulated by noxious chemicals, how their function differs in different organs, and how the transfer of signals occurs. As soon as signal transfer has occurred, however, the actions are, as always, controlled by the spinal cord or the brain.

The signal for mechanical pressure and its pain is sensed in an analogous fashion. There are a variety of nerve fibers that sense pressure; in fact, some of the nerve fibers sensing temperature also sense pressure (Figure 6.3) [7–9]. There are different sensory corpuscles present in human hairy skin [9]. Low threshold mechanoreceptors contact with epithelial Merkel cells or Schwann-like cells forming Merkel cell neurite complexes (slowly adapting), Meissner corpuscles (rapidly adapting), Pacinian corpuscles (rapidly adapting), and Ruffini endings (slowly adapting). Hairs contain several nerve endings; occasionally, hairs have associated Merkel cells, Ruffini and even Pacinian corpuscles. These sensory nerve cells are concentrated in "touch domes", slight dermal protrusions or bumps [10]. Touch domes also include Merkel cells. Merkel cells are specialized cells in the epidermis that sense light touch via an indirect pathway, i. e., the Merkel cells release a chemical that only then will activate a nerve cell (Figure 6.1). Merkel cells react to light touch slowly, since the signal needs to diffuse first to neurons before it can be detected.

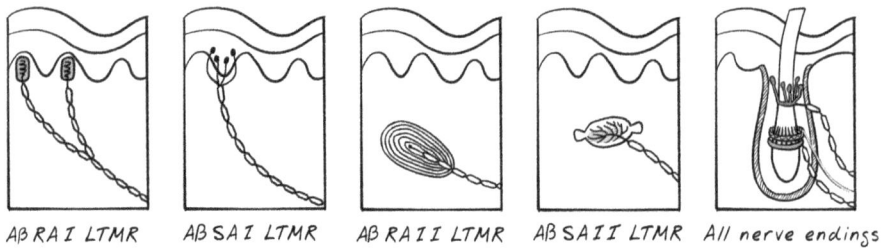

Aβ RA I LTMR   Aβ SA I LTMR   Aβ RA II LTMR   Aβ SA II LTMR   All nerve endings

Figure 6.3: Scheme of the different sensory corpuscles present in human hairy skin. Low threshold mechanoreceptors contact with epithelial Merkel cells or Schwann-like cells forming Merkel cell neurite complexes (slowly adapting), Meissner corpuscles (rapidly adapting), Pacinian corpuscles (rapidly adapting), and Ruffini endings (slowly adapting). Hairs contain several nerve endings; occasionally, hairs have associated Merkel cells, Ruffini and even Pacinian corpuscles. (adapted from [9]).

As with temperature, extreme pressure will be sensed as pain. When pain is sensed it is highly regulated. There are additional receptors that can up- or down-regulate the pain sensation; some of these receptors are also located at the nerve ending of the sensory neuron [11]. One example is the endothelin-1 receptor (Figure 6.4). Up- or down-regulation usually occurs via a signal transduction pathway and has several points where a variety of different chemicals can modulate transduction, either directly or allosterically. Some of these are redundant pathways. One such regulation is the increasing pain you feel when pressure is applied for a long time, but also the reduction of pain after an initial, short-term stimulus. Additionally, heat might modulate pain perception as well, which might explain why some sensor elements are activated by both.

One of the multifunctional sensor elements we have already seen: Transient receptor potential ion channel TRPV4, which is one of the channels for sensing heat (see Figure 6.2, Table 6.1) [12]. It is one of the channels opening at higher temperature that is also linked to pain and activation by chemicals (specifically capsaicin). It is widely distributed. TRPV4 mutations in humans lead to impaired bone development, motor dysfunction, sensory loss of the combined thermal/pain sensation, abnormal sensation of osmolarity and blood salt concentration, hearing impairment, bladder dysfunction, and airway dysfunction [12]. Mutations in animals suggest even more functions, including the barrier function of skin, insulin secretion, and liver function—there might be more. The connection of all of these dysfunctions to the heat and mechanical sensor are still under debate. TRPV4 works with several signal transduction pathways; these pathways give some hints as to how this might be possible (Figure 6.5). Additionally, there are different mechanical stimuli that the ion channel reacts to differently: membrane stretching, membrane press or pull, shear stress, and swelling of the cell in hypotonic solutions.

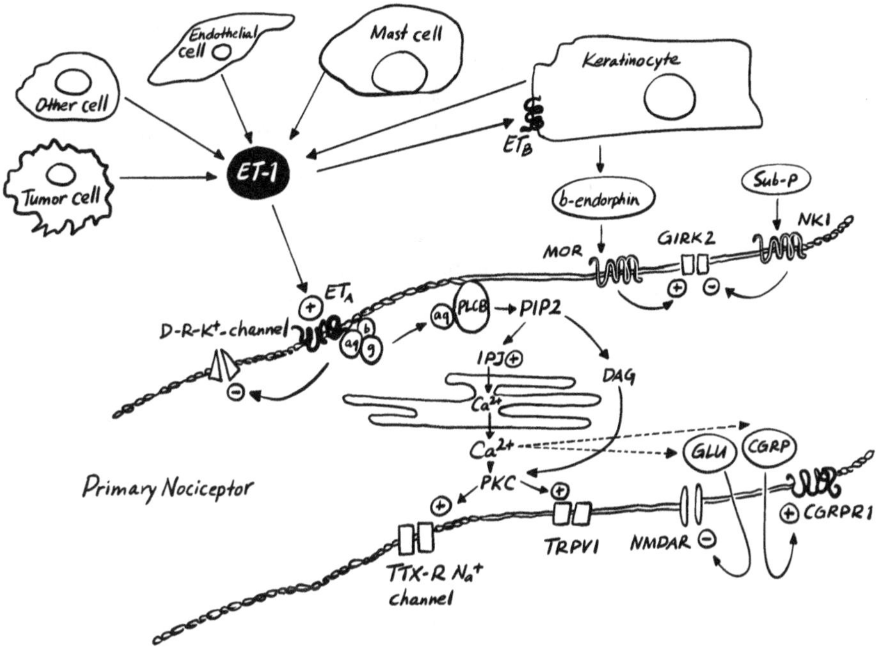

Figure 6.4: A specific pain receptor, the endothelin-1 receptor is up- or down-regulating via a signal transduction pathway (adapted from [11]).

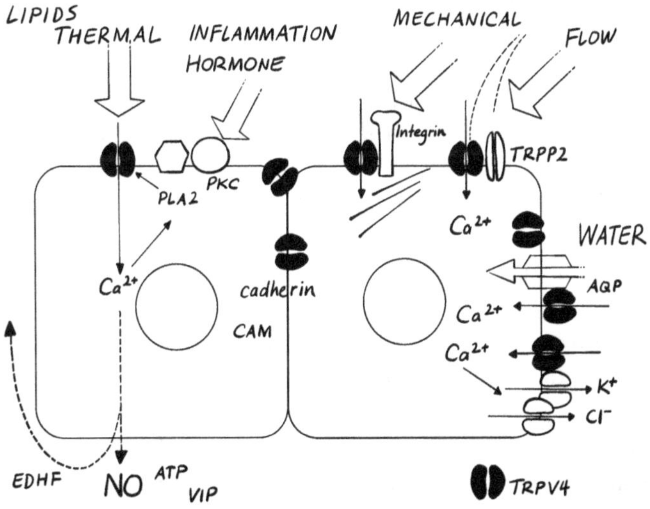

Figure 6.5: Different signal transduction pathways of TRPV4, resulting in this heat and mechanical sensor being connected to other functions of skin, including its barrier function, as well as functions all across the body, including bone development, motor dysfunction, abnormal sensation of osmolarity and blood salt concentration, hearing impairment, bladder dysfunction, and airway dysfunction (adapted from [12]).

One of the dysfunctions, hearing impairment can be explained more easily: TRPV4 is one of the ion channels in the hair cells in the cochlea that we discussed in Chapter 5.1. So, these ion channels are not only part of the sensation of touch, they also detect pressure in the hair cells, and thus are part of the indirect signal transduction in hearing [12].

These ion channels are part of neurons that are part of skin (or another organ), which means that the pressure the ion channels feels first has to be transduced through a layer of skin. Therefore, there will always be a filtering effect by the mechanical properties of the skin or organ. The effect of the tissue has mostly been studied in animals, and in a few cases, it has been found that the tissue can amplify the signal, instead of the more common dampening effect that occurs [13].

Let us look at the indirect method of the mechano-sensation, which are the Merkel cells (Figure 6.1) [14]. Merkel cells react to light touch, but they react slowly and continue to fire with long stimulation with action potentials across their cell membrane. Their firing pattern is irregular. There is likely a link between Merkel cells and nerve terminals that use glutamate as their neurotransmitter. There seem to also be Merkel cells without a connection to nerve cells; at that point, the signal might be an exocytosed chemical diffusing into the environment. So far, though, there is no evidence of exocytosis.

Besides being a large area with different sensing neurons, skin's other important function is to protect against infection and injury. How does skin fulfill all of these different functions simultaneously? Its structure plays a key role here (Figure 6.1) [15]. The top layer, the epidermis, is the tough protective layer. It consists of dead keratinocyte cells that are stacked like bricks to reduce pathways for disease agents to enter the body. Additionally, between the dead cells, there is an extra-cellular matrix (ECM), a mixture of fibers (collagen, elastin), and amorphous matrix (proteoglycans, cell-binding adhesive glycoproteins, solutes, water) that acts as a mechanical support and surface to anchor cells. The ECM also determines cell orientation, controls cell growth and differentiation via signals, and scaffolds the three-dimensional structure for optimal tissue environment and signal transport.

When an injury occurs, skin has the ability to heal itself also due to its complex structure [15, 16]. Blood clotting is the first step in the process (Figure 6.6), a signal cascade that creates a thrombus or clot. The thrombus is the initial matrix and initiates inflammation, which starts the healing process. The matrix of the clot is reorganized while signals recruit fibroblasts, which themselves release more signals. The reorganizing matrix also helps with building a specific structure by providing attachment points for migrating cells. Inflammation occurs via several pathways, each of which is a signal cascade with a variety of regulation points (Table 6.2) [15, 17]. Inflammation also initiates the innate and adaptive immune system to combat any disease-carrying bacteria and viruses that might have come in through the breach in the skin. Therefore, the automatic healing system is a complex, overlapping system that is highly controlled and adaptive to the actual injury and specific environment. Not only will blood clotting quickly stop the bleeding, but the basic fibers will be reorganized so that cells can attach to them in an organized fashion, recreating an organized tissue with blood vessels,

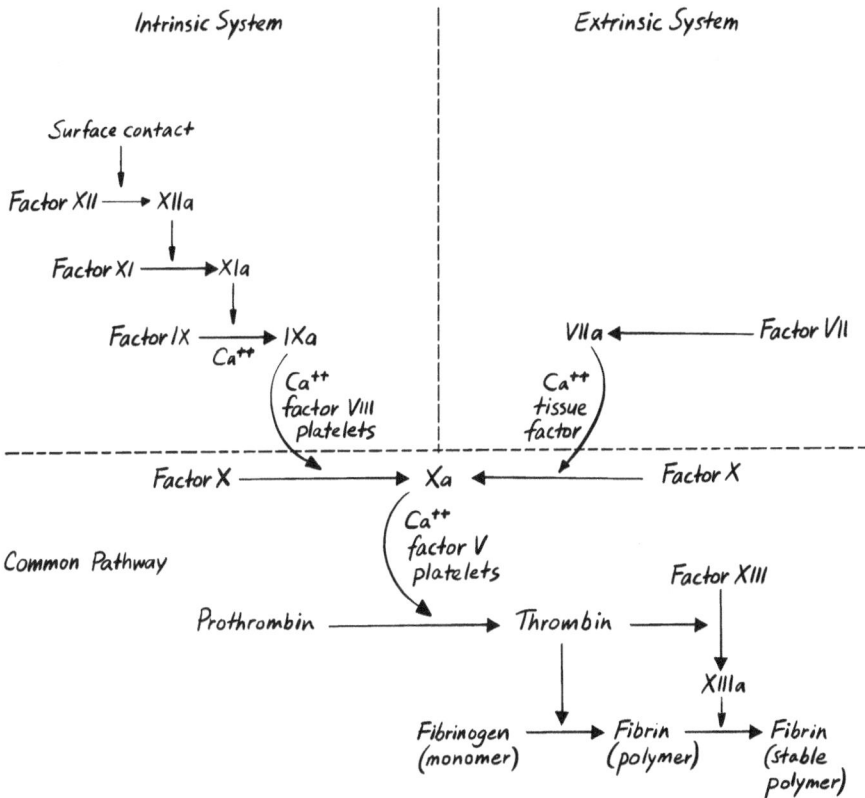

Figure 6.6: The blood-clotting cascade that creates the blood clot, the first step in healing a wound (adapted from [15]).

Table 6.2: The different pathways of inflammation, the next step in the healing process of skin after the bleeding is stopped [15].

Mediators	Examples
Vasoactive Agents	Histamines, serotonin, endothelin, thromboxane $\alpha 2$
Plasma Proteases	Bradykinin, complement system C3a, C5a, C3b, C5b-C9, fibrin degradation products, tissue plasminogen activator
Leukotrienes	Leukotriene B4
Lysosomal Proteases	Collagenase, elastase
Oxygen-Derived Free Radicals	H_2O_2, superoxide anion
Platelet Activating Factors	Cell membrane lipids
Cytokines	Interleukin-1, tumor necrosis factor
Growth Factors	Platelet-derived growth factor, transforming growth factor, epithelial growth factor

nerves, and other specialized cells for all the functions of skin. Over time, the system no longer works perfectly, thus resulting in scars. In addition, when all layers of skin have been destroyed by an injury (e. g., in the case of a third-degree burn), the scars still allow for survival but a few of the functions of skin will not be perfectly recreated (such as cooling or sensing). Even so, functional skin will be regrown automatically, in spite of the possibility of wound and other complicating factors.

Let us summarize how our human temperature and touch sensors work: Both senses are imbedded in nerves that include special ion channels that open in response to a heat or touch signal. Temperature-sensitive channels open using a thermodynamics-based mechanism whose details are not known yet. In the case of the mechano-sensitive ion channels, the stretching of the skin-environment physically opens the channel, thus creating a membrane potential that can directly or indirectly initiate an action potential in the next nerve cell. This starts a path to the brain, where analysis takes place, which includes the possible sensing of "pain" or either of these senses (in fact, pain can also occur in hearing; however, this only takes place in rather extreme circumstances). Some of the same channels that react to temperature or pressure also react to chemicals that create a burning or cooling sensation. The sense of temperature or pressure leads to some automatic avoidance movements or conscious decisions and behaviors.

At the same time, the organ with these sensing neurons happens to be skin, the biggest organ that has additional functions, such as maintaining homeostasis, temperature control of the body, protection from injury, bacteria and viruses, and Vitamin D production. It is involved in the large majority of adaptive behaviors of the body (which is how skin maintains homeostasis), since it is the organ that borders the outside environment. In other words, skin is many different sensors embedded in a "smart", adaptive matrix. Could we use molecules and methods of the human senses to make an artificial, smart, sensing material with similar function(s)?

6.2 Chemical, Thermal, and Pressure Sensors Using Biological Cells, Molecules, and Methods

Remember how the two main sensors in skin worked: For temperature, an ion channel in a nerve cell opened based on temperature via a (so far) unclear mechanism. No literature was found about using the nerve cells or ion channels in artificial temperature sensors. This is likely due to the fact that a) the channel mechanism is not known, and thus difficult to control outside its native environment and b) there are other highly developed temperature sensors already on the market, thermometers, with even more in development, even on a very small scale.

Thermometers on a large scale use liquids that change their volume with temperature. On the microscale (i. e., when they are small enough that they fit on a computer chip), most temperature sensors are thermocouples [18]. A thermocouple combines two different metals (often two small wires). When heated, a small voltage will be generated, and the relationship between voltage and temperature can be calibrated. An analogous method can be used with semiconductors in P-N junctions. Since P-N junctions are the basis of several chip building blocks (e. g., for transistors), with the right circuit design the circuit can function as intended and measure the temperature at the same time. These measurements are digital, and thus easy to incorporate into an analysis system as they are in the skin/brain connection.

Pressure-measurements are common on the microscale as well. In that case, piezoelectric materials are used as already seen in Chapter 5.3 [19–21]. There are a variety of piezoelectric materials such as perovskites (e. g., [22] and references therein) for a variety of applications, such as energy harvesting (e. g., [23] and references therein), high-temperature transducers and actuators (e. g., [24] and references therein), and tissue regeneration (e. g., [25] and references therein), and more are being developed (e. g., [26]). Due to the availability of a large amount of piezoelectric materials, there was no need to use an ion-channel that opens when the surrounding membrane stretches— a channel that is likely to be difficult to maintain and control in a nonnative environment.

There are a large number of different "smart", or environment-responsive, materials that have been reported, some of them containing polysaccharides or proteins. Some of these examples will be discussed in the following biomimetic skin section, 6.3.

6.3 Biomimetic Skin

The function of skin that has been most often mimicked is touch. This could include simple piezoelectric materials sensing pressure, as mentioned in the preceding section. But touch is a lot more complex: Touch also senses surface texture and helps with gripping. And gripping is automatically adjusted based on the strength of the material and the weight of the sample. This is important for artificial limbs and robots.

An example for a system that mimics the structure of skin to develop a touch sensor with a sensitivity approaching that of human skin is discussed here [27] (Figure 6.7). This sensor includes several design elements of skin: The surface contains ridges that are

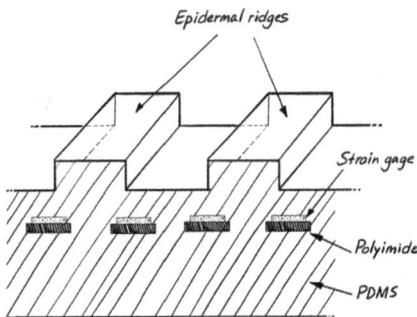

Figure 6.7: A touch sensor that mimics the structure of skin to develop a sensitivity approaching that of human skin (adapted from [27]).

mechanically stiffer than the rest of the surface, as the epidermis does, and the pressure sensing elements are placed in the same position relative to the ridges as the human Merkel discs in skin. Additionally, the strain sensors show different strain values based on the direction of the stress, which with the right algorithm can be translated into the sensation of surface properties. The "skin" material is easily deformable, allowing for strain transduction. Research results demonstrate that ridges enhance sensitivity and add information about the three-dimensional structure of the surface as human touch is able to do.

To mimic skin, pressure-sensing needs to be scaled up to larger areas. A sensor based on carbon nanotubes is attempting this (Figure 6.8) [28]. The sensor contains single-walled CNTs in a cellulose layer. This layer can be spray-coated, allowing for facile fabrication for larger sensors. Additionally, the electrodes that will transduce the pressure into an electrical signal via resistive measurements can be ink-jet printed, which can also be scaled-up easily.

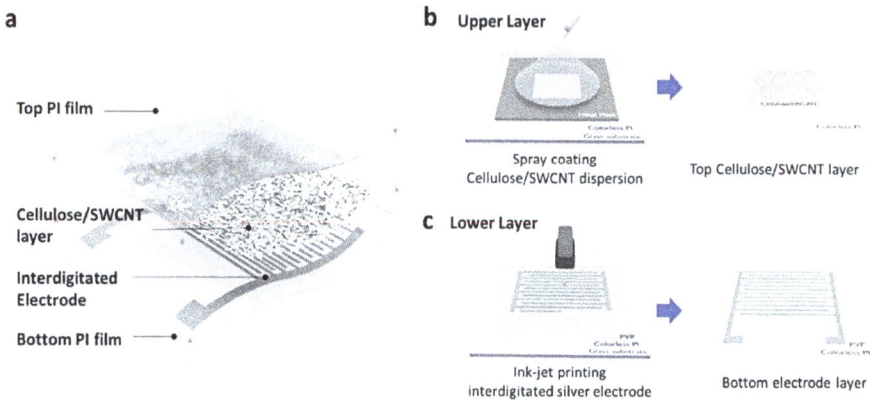

a

Top PI film

Cellulose/SWCNT layer

Interdigitated Electrode

Bottom PI film

b Upper Layer

Spray coating
Cellulose/SWCNT dispersion

Top Cellulose/SWCNT layer

c Lower Layer

Ink-jet printing
interdigitated silver electrode

Bottom electrode layer

Figure 6.8: Piezoresistive flexible cellulose pressure sensor containing single-walled CNTs (with permission from [28]). (a) Schematic of the layers of the sensor. (b) Schematic of the fabrication process of spray coating the CNT/cellulose sensing film onto the protective polyimide film. (c) Schematic of inkjet printing the interdigitated silver electrode.

Another group is connecting temperature sensing with actuation: at a low temperature, the material sensor will bend [29]. Adding electrodes to an asymmetric graphene-coated paper becomes a flexible piezoresistive strain sensor. When coating this film with a porous hydrophilic cellulose fiber network, the adding/removing of water at different temperatures makes the film bend, thus adding actuation to the sensor film.

Layering two sensor films on top of each other created a film that can measure both temperature and pressure (Figure 6.9) [30]. A capacitive pressure sensor array in a dielectric elastomer is placed between two electrodes. On top of that, a layer with a tem-

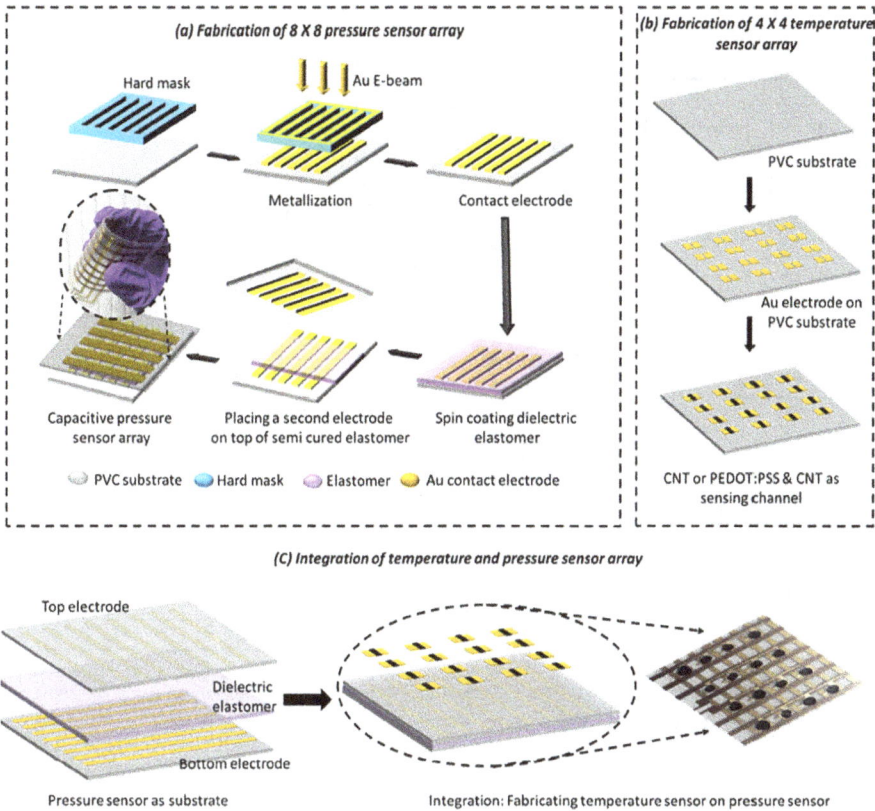

Figure 6.9: Fabrication scheme of a multifunctional layered sensor (with permission from [30]): (a) Fabrication of a capacitive pressure sensor array with a dielectric elastomer placed between electrodes; (b) fabrication of temperature sensor array by drop casting CNTs over contact electrodes; (c) integration of temperature and pressure sensor layers.

perature sensor using CNTs drop casted over contact electrodes is placed. Those two layers are now a film that could, for example, be used as the "skin" of robots.

In a different approach, the combination and management of a large amount of pressure sensors has been investigated [31] (Figure 6.10). Here, an integrated circuit is designed to manage a large amount of sensors on a chip, concentrating on how all of the sensors can be packaged, networked, and their signal transduced on a chip. It might be interesting to combine that management with the earlier sensor designs, since in skin a large amount of pressure and other sensors need to be managed and analyzed.

In skin, the sensing of temperature or pressure leads to a related action in the body, which could be an adjustment to maintain homeostasis, a movement, or an action the brain thought of. This responsiveness to the environment has been mimicked in materials in various ways with a variety of polymeric structures, called "smart" materials ([32] and references therein) (Figure 6.11, Figure 6.12). The stimuli for response are most

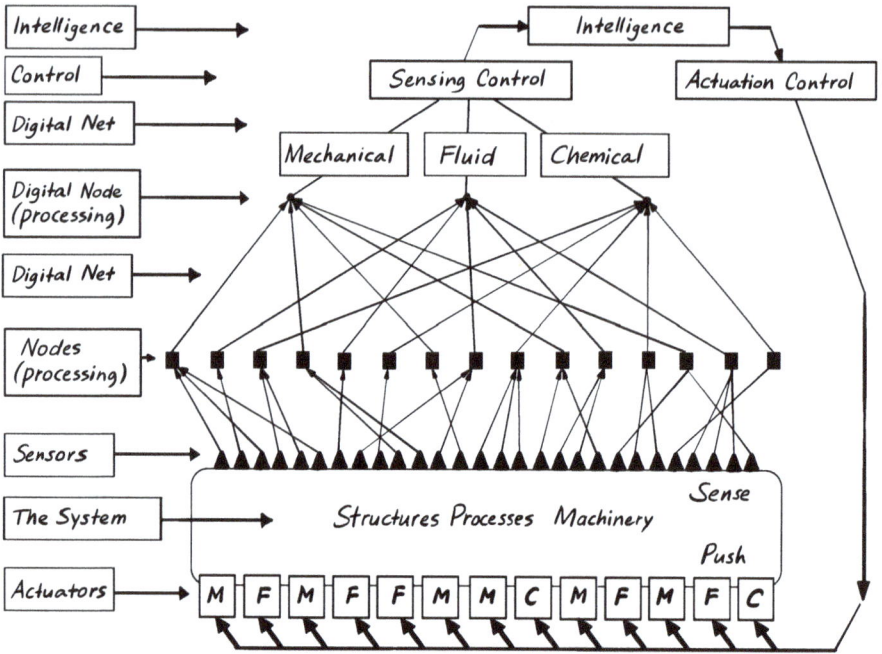

Figure 6.10: An integrated circuit manages and analyzes a large amount of pressure sensors. The sensors are networked and transduced in a similar complexity as seen in skin (adapted from [31]).

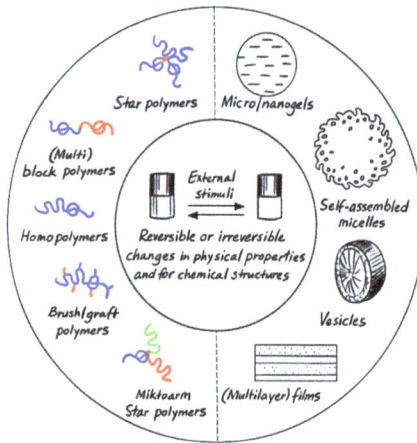

Figure 6.11: "Smart materials" can be made from a variety of different structures (adapted from [32]).

commonly temperature and pH, but they could also be a specific binding event. The triggered action can be a conformation or phase change, a polymerization or polymer degradation, or a color change due to a change in conjugation length. The "read-out" after transduction could be the change in color, or could involve fluorescent molecules that are sensitive to different pHs or phase states. These responsive materials are often

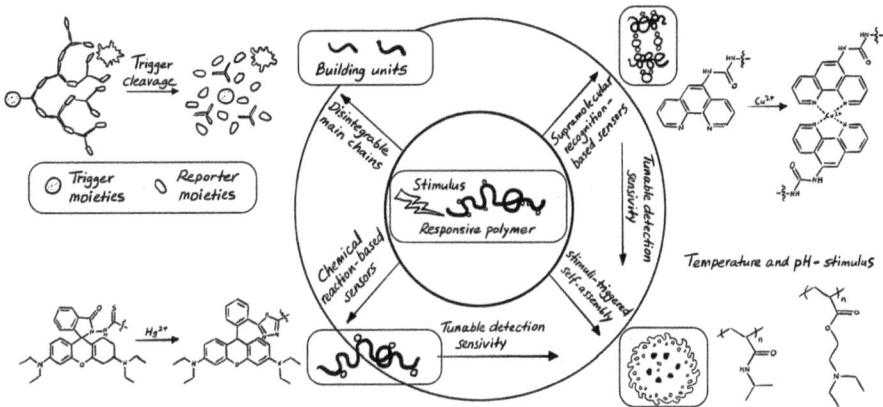

Figure 6.12: "Smart materials" react to a variety of stimuli with several different responses (adapted from [32]).

made from biological polymers for medical applications to make them more biocompatible ([33] and references therein).

A lot of smart materials were developed for drug delivery ([34] and references therein) (Figure 6.13). Generally, these are polymeric capsules that rupture upon a specific signal. Therefore, they release more specifically next to/inside of a target tissue/cell, and thus reduce the side effects of drugs that are caused by the drug's distribution into healthy tissues and cells.

To make drug delivery even more specific, the body's system of molecular recognition and specific transport can be used. One such method is to use the cell's host-guest interactions to only release in the presence of a specific cell marker or metabolite ([35] and references therein). This specific recognition is then transduced to either a read-out (when used in sensors) or is a trigger for release (when used in drug delivery), sometimes even in self-healing applications.

Another method to use the body's specificity is to make the action responsive to an enzyme reaction ([36] and references therein). Here, the trigger includes an enzyme that only acts in a specific cell or disease state. In enzyme-responsive systems, it is possible to also diagnose a disease and only act after diagnosis has occurred, if the enzyme happens to be a marker for a specific disease.

The other important property of skin, self-healing, has also been mimicked; a large number of different approaches for different materials and applications have been reported (see recent review and references therein [37]). Only a small number of examples are highlighted here.

The first generation of self-healing materials was designed for structural materials [38, 39] (Figure 6.14). Basically, the polymeric material contains capsules with monomer (labeled "healing agent" in the graph) and in its matrix also a small amount of catalyst. When a crack develops, it opens a capsule with monomer, which is a liquid that will

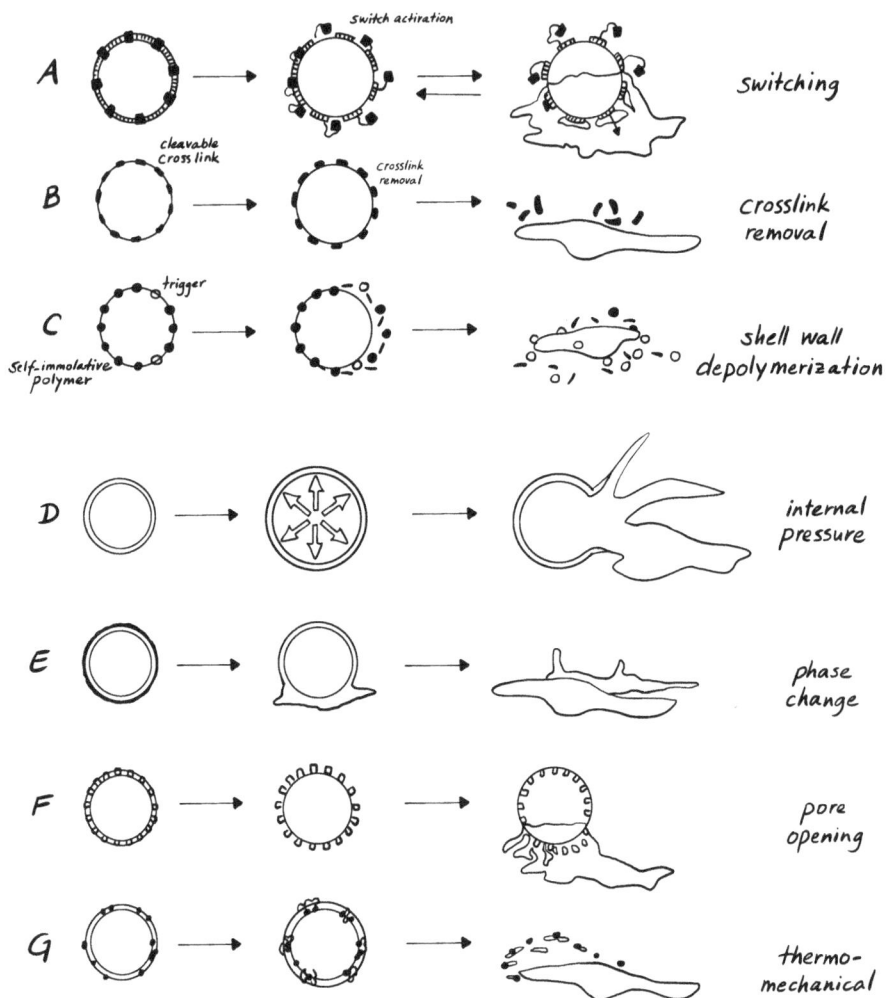

Figure 6.13: Drug delivery with smart materials (adapted from [34]).

start flowing. It will flow past a catalyst, which will start polymerization, forming new polymer in the crack, thus "healing" the matrix. In a lot of cases, healing additionally requires heat or light, since most polymerization initiators require heat or light for activation. A lot of research has been performed to optimize the outer capsule membrane, the distribution of catalyst and capsules in the matrix, the kinetics of the healing polymerization, and the strength recovery for the material [38, 40]. Analogous work has been done for supramolecular systems, i. e., systems based on intermolecular forces instead of covalent bonds [41], as well as for nanostructured materials [42].

For biomaterials that will be used in medicine, it is important that the material used is a hydrogel so that it can take advantage of water-based transport. At the same time,

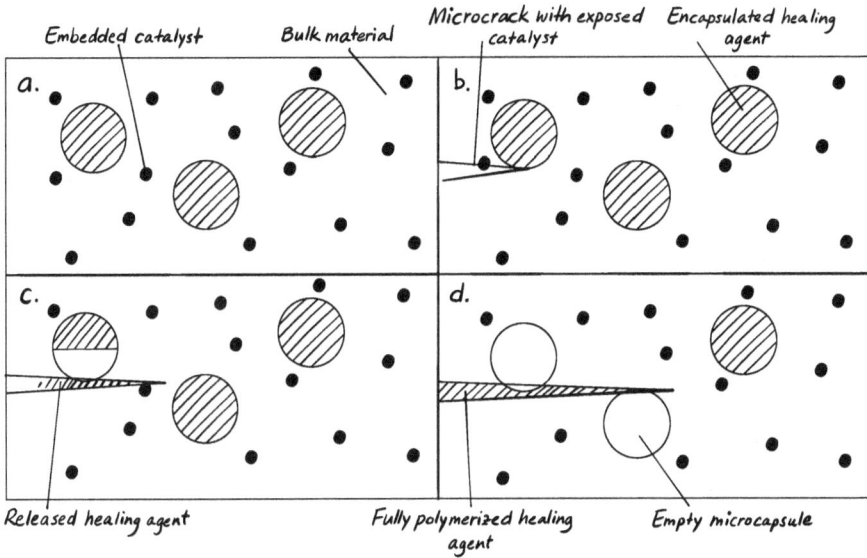

Figure 6.14: Approaches for self-healing structural materials (adapted from [39]).

the healing process needs to operate at physiological conditions, i. e., 37 °C and a neutral pH. An example of such a system is zwitterionic hydrogels [43] (Figure 6.15). Basically, the healing process is based on charge attraction; charged polymers are flexible and have a large driving force to recombine with the opposite charge. Therefore, healing is automatic and fast. It is important, though, to create a material that is stable at a temperature above body temperature.

So far, the important properties of skin, its function as a platform for sensors, its responsiveness to the environment, and its self-healing abilities, have been investigated separately. But is there a material that is as complex in its structure and functions as skin that can do all of this at the same time, as skin can? The short answer is: not yet. There are a lot of examples of biocompatible hydrogels in the tissue engineering literature (e. g., [44] and references therein) that encourage the redevelopment of the tissue structure. Water and chemical transport in hydrogels [45–47] and the exact mechanical properties [48] have been investigated as well. However, these are just platforms that the other properties (sensing and responsiveness to the environment) can be built upon. In addition, as materials go, scaffold hydrogels are not yet self-healing.

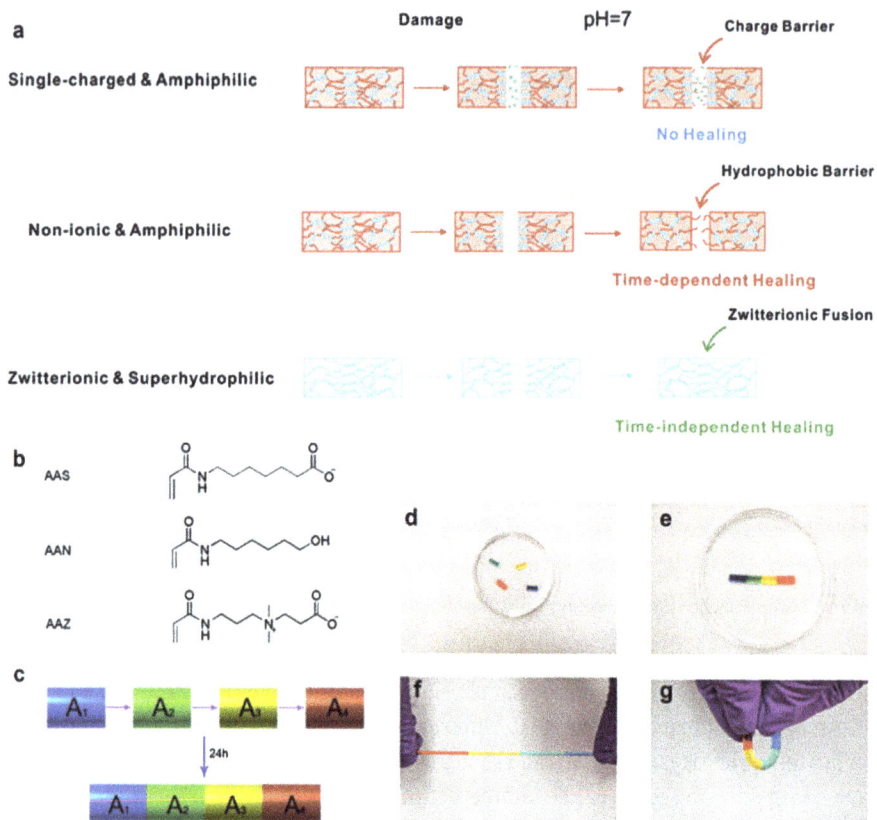

Figure 6.15: A self-healing hydrogel for medical applications. This is a mimic for skin without the sensing and other functions of skin [43]. (a) Different healing behaviors of single-charged, nonionic and zwitterionic materials. (b) Chemical structures of the different materials used. (c) How the different blocks combine to form the block hydrogel. (d) Individual hydrogel building blocks dyed with different colors. (e) A completed block hydrogel. (f) The block hydrogel upon stretching. (g) The block hydrogel upon bending.

6.4 Summary and the Bigger Picture

To summarize, skin is the largest and most complex organ with the most different functions and the largest amount of different sensors in the human body. The sensing mechanisms have not yet been fully worked out in detail, neither for temperature nor for pressure. Additionally, these sensors are also part of chemical sensing and part of the sensation of pain; and all of this sensing is so sensitive that the details of surface properties can be distinguished in a three-dimensional manner. Additionally, these sensors are built into a complex structure that also acts to maintain homeostasis and self-heals when the barrier to the outside environment is breached. Some of these functions have been mimicked in materials but generally separately. There has been progress of building sensor arrays on a large scale, including progress of addressing each of the sensors separately, as the human body does as well. No material has so far been reported with the same functionality as skin.

Bibliography

[1]   Sharma A. Medical Biochemistry: Molecules to Disease. iBooks, https://itunes.apple.com/us/book/medical-biochemistry-molecules/id1119168051?mt=11. 2016.
[2]   Dhaka A, Viswanath V, Patapoutian A. Annual Review of Neuroscience. 2006;29:136–161.
[3]   Nagy I, Santha P, Jancso G, Urban L. European Journal of Pharmacology. 2004;500:351–369.
[4]   Frey NV, Gavrin JR. Thomas Hematopoietoc Cell Transplantation. 2009;1570–1588.
[5]   Nakamura K, Morrison SF. Nature Neuroscience. 2008;11:62–71.
[6]   Cannon B, Nedergaard J. Physiological Reviews. 2004;84.
[7]   Schmidt R, Schmelz M, Torebjork HE, Handwerker HO. Neuroscience. 2009;98:793–800.
[8]   Marshall KL, Lumpkin EA. In: López-Larrea C, editor. Sensing in Nature. Landes Bioscience and Springer Science+Business Media; 2012.
[9]   Cobo R, García-Piqueras J, García-Mesa Y, Feito J, García-Suárez O, Vega JA. International Journal of Molecular Sciences. 2020;21:6221. https://doi.org/10.3390/ijms21176221.
[10] Reinisch CM, Tschachler E. Annals of Neurology. 2005;58:88–95.
[11] Khodorova A, Montmayeur JP, Strichartz G. The Journal of Pain. 2009;10:4–28.
[12] Suzuki M. In: Mechanically gated channels and their regulation. 6 2012;103–157.
[13] Katta S, Krieg M, Goodman MB. Annual Review of Cell and Developmental Biology. 2015;31:347–371.
[14] Halata Z, Grim M, Baumann KI. Expert Review of Dermatology. 2010;5:109–116.
[15] Ratner BD. Biomaterials Science: An Introduction to Materials in Medicine. San Diego: Academic Press; 1996.
[16] Garg HG, Longaker MT. Scarless Wound Healing. New York, NY: Marcel Dekker, Inc.; 2000.
[17] Clark RAF. The Molecular and Cellular Biology of Wound Repair. 2nd ed. New York, NY: Kluwer Academic Publishers; 1996.
[18] Ramsden E. Sensor Magazine, 2000. http://www.sensorsmag.com/components/temperature-measurement.
[19] Lee HS, Chung J, Hwang GT, Jeong CK, Jung Y, Kwak JH, Kang H, Byun M, Kim WD, Hur S, Oh SH, Lee KJ. Advanced Functional Materials. 2014;24:6914–6921.
[20] Hwang GT, Byun M, Jeong CK, Lee KJ. Advanced Healthcare Materials. 2015;4:646–658.
[21] Lee WS, Lee SS. Sensors and Actuators A. 2008;144:367–373.
[22] Uchino K. Science and Technology of Advanced Materials. 2015;16:046001-046001–046001-046016.
[23] Eklund P, Kerdsongpanya S, Alling B. Journal of Materials Chemistry C: Materials for Optical and Electronic Devices. 2016;4:3905–3914.
[24] Stevenson T, Martin DG, Cowin PI, Blumfield A, Bell AJ, Comyn TP, Weaver PM. Journal of Materials Science: Materials in Electronics. 2015;26:9256–9267.
[25] Rajabi AH, Jaffe M, Arinzeh TL. Acta Biomaterialia. 2015;24:12–23.
[26] Batth A, Mueller A, Rakesh L, Mellinger A. In: Proceedings, Conference on Electrical Insulation and Dielectric Phenomena (CEIDP). 2012.
[27] Zhang Y, Miki N. Journal of Micromechanics and Microengineering. 2010;20:085012-085011–085012-085017.
[28] Kim D, Lee DK, Yoon J, Hahm D, Lee B, Oh E, Kim G, Seo J, Kim H, Hong Y. ACS Applied Materials Interfaces. 2021;13:53111–53119.
[29] Hu Y, Qi K, Chang L, Liu J, Yang L, Huang M, Wu G, Lu P, Chen W, Wu Y. Journal of Materials Chemistry C. 2019;7:6879.

[30]   Kumaresan Y, Ozioko O, Dahiya R. IEEE Sensors Journal. 2021;21:26243. https://doi.org/10.1109/
        JSEN.2021.3055458 (open access: https://creativecommons.org/licenses/by/4.0/).
[31]   Jacobsen SC, Maclean BJ, Whitaker M, Olivier M, Mladejovsky MG. Proceedings of SPIE.
        1999;3673:19–32.
[32]   Hu J, Liu S. Macromolecules. 2010;43:8315–8330.
[33]   Randolph LM, Chien MP, Gianneschi NC. Chemical Science. 2012;3:1363–1380.
[34]   Esser-Kahn AP, Odom SA, Sottos NR, White SR, Moore JS. Macromolecules. 2011;44:5539–5553.
[35]   Hu J, Liu S. Accounts of Chemical Research. 2014;47:2084–2095.
[36]   Ulijn RV. Journal of Materials Chemistry. 2006;16:2217–2225.
[37]   Speck T, Bauer G, Flues F, Oelker K, Rampf M, Schuessele AM, Von Tapavicza M, Bertling J,
        Luchsinger R, Nellesen A, Schmidt AM, Muelhaupt R, Speck O. Materials Design Inspired by Nature.
        2013;4:359–389.
[38]   Ollier R, Rodriguez E, Alvarez V. Advances in Materials Science Research. 2013;16:121–148.
[39]   Brochu ABW, Craig SL, Reichert WM. Journal of Biomedical Materials Research A. 2011;96:492–506.
[40]   Wu G, An J, Tang XZ, Xiang Y, Yang J. Advanced Functional Materials. 2014;24:6751–6761.
[41]   van Gemert GML, Peeters JW, Söntjens SHM, Janssen SM, Bosman AW. Macromolecular Chemistry and
        Physics. 2012;213:234–242.
[42]   Kötteritzsch J, Schubert JS, Hager MD. Nanotechnology Review. 2013;2:699–723.
[43]   Bai T, Liu S, Sun F, Sinclair A, Zhang L, Shao Q, Jiang S. Biomaterials. 2014;35:3926–3933.
[44]   Gelain F. International Journal of Nanomedicine. 2008;3:415–424.
[45]   Hoffmann AS. Advanced Drug Delivery Reviews. 2002;43:3–12.
[46]   Hatakeyama T, Yatakeyama H. Thermal Properties of Green Polymers and Biocomposites. Dordrecht,
        NL: Kluwer Academic Publishers; 2010.
[47]   Juris SI, Mueller A, Smith BTL, Johnston S, Kross RD. Journal of Biomaterials and Nanobiotechnology.
        2011;2:216–225.
[48]   Saxena S, Hansen CE, Lyon LA. Accounts of Chemical Research. 2014;47:2426–2434.

7 Future Developments

This book is about human senses and movement, how they function on the nanoscale, and how they can be mimicked on the nanoscale by technology. As we have seen, human and human-made sensors always have several elements: the sensing or detection element, the transducer, possibly an amplifier, and an analysis element.

The human sensing elements, eyes, nose, ears, and the mechanical and temperature sensors in skin, are connected to receptors or ion channels as transducers that change the original signal into an electrical signal. This initiates an action potential, which is sent to the brain for analysis and possible action/reaction based on need or memory/experience. In technology, sensors work in very much the same way. The sensing element is translated into an electrical signal that is sent to a computer for analysis and recognition. Nowadays, computers might even remember the signal or a signal pattern and learn from that "experience".

Human motion is based on two stiff, molecular molecules, one that is the "street", and the other that uses energy to "walk" on that street. A large amount of these molecules combine in a muscle. The duration and the strength of muscle contraction can be controlled in this system by controlling these molecules. In nanotechnology, these molecules have been successfully used to transport nanosized cargo. The system can now in a few cases be automated—with the exception of loading the cargo onto the motor protein. Initial work has been successful in scaling up movement by self-assembly, but the problem of large-scale and long-duration movement has not yet been solved. Mimicking motion with other chemicals is still in its infancy, since there has not been an effective street/motor connection developed with the exception of rotaxane-based molecules. But those molecules limit motion to a few nanometers and do not allow for continuous, linear motion. Some self-assembled systems using DNA or vesicles are trying to create controlled larger-scale movement in tubes. Rotary motion is easier with chemicals, but it is difficult to connect these chemicals to a controlled energy source.

In the eyes, the human photoreceptors, detection occurs via a change in the three-dimensional structure of a molecule. The change in shape initiates a signal cascade that amplifies the signal and transfers the signal to the brain. Using the actual molecules in the process for nanosized photosensors is difficult, and has only rarely been successful. There is a large variety of nanotechnological photosensors. They range from chemical or electrical sensors that measure a change in light absorption to fluorescence sensors where a chemical or current is reported as a color change. What is less common is that sensors are connected to amplification and analysis within one system. Phototransistors are trying to start to achieve that integration. Another function of human system has been mimicked, however—the constant movement of the eyes to create a more accurate picture and to help with the analysis of it. Combining all of these properties would make for a much more powerful photosensor in the future.

Smell and taste have been mimicked most often, and chemical sensing methods are the furthest developed among approaches to replicating theses senses. These methods

https://doi.org/10.1515/9783110779196-007

often combine multiple sensing elements with a transducer and an analysis element (generally a form of pattern recognition). The specificity of binding/sensing is currently still the biggest problem that must be overcome.

Hearing on the molecular scale is actually rather complicated: Vibrations in air are converted into mechanical vibrations in a membrane via various steps and membranes and eventually are transduced into electricity via the bending of fiber that mechanically opens an ion channel. The advantages of this complicated set-up can be seen in the human ear's excellent sensitivity, since wavelengths are detected separately and only recombined during the analysis-step in the brain. Since this system is so complicated, there has not been much research into using these membranes or sensors in nanotechnology, but so far largely only investigations into the system itself. On the technology side of things, microphones and receivers are commonplace on any scale. However, none of these has so far achieved the efficiency of the basilar membrane. Acoustic sensors, though, are starting to be used in different applications, such as sensing flow.

Skin is the largest and most complex human organ with the widest variety of functions. The temperature or pressure sensors are specialized neurons and contain 3D information. Additionally, these sensors are also part of chemical and pain sensing. These sensors are built into a complex structure that self-heals and maintains homeostasis. Some of these functions have been mimicked in materials, but only one function at a time has been successfully reproduced. Layered sensor films are starting to integrate sensors on a larger scale.

In humans, all of the senses are combined into one structure, our bodies. All of the sensor elements are analyzed by the same entity, the brain. Analysis here does not only imply the recognition of a specific shape such as a rose, but it also implies the creation of a connection between the rose and the feeling of love, also using memory in the process. Robots are the technological answer to such a complex system and they use quite a few senses. However, no robot has yet used senses that are as equally fine-tuned as the human senses on the molecular scale in combination with rational thought and complex action based on that thought.

Index

https://doi.org/10.1515/9783110779196-008

www.ingramcontent.com/pod-product-compliance
Lightning Source LLC
Chambersburg PA
CBHW081542220326
41598CB00036B/6534